H. Werner, Discriminator-Algebras

STUDIEN ZUR ALGEBRA UND IHRE ANWENDUNGEN

Im Auftrage des
Zentralinstituts für Mathematik und Mechanik
der Akademie der Wissenschaften der DDR
herausgegeben von L. Budach und H.-J. Hoehnke

BAND 6

H. WERNER

DISCRIMINATOR-ALGEBRAS

ALGEBRAIC REPRESENTATION AND MODEL THEORETIC PROPERTIES

HEINRICH WERNER

DISCRIMINATOR-ALGEBRAS

ALGEBRAIC REPRESENTATION AND MODEL THEORETIC PROPERTIES

AKADEMIE-VERLAG · BERLIN
1978

Zentralinstitut für Mathematik und Mechanik, 108 Berlin, Mohrenstr. 39
Erschienen im Akademie-Verlag, 108 Berlin, Leipziger Straße 3–4

Lizenznummer: 202 · 100/423/78
Gesamtherstellung: VEB Druckerei „Thomas Müntzer“, 582 Bad Langensalza
Bestellnummer: 762 519 8 (2174/6) · LSV 1025
Printed in GDR
DDR 18,– M

Contents

Algebraic methods became an important tool in the mathematical logic after the algebraization of the propositional calculus by GEORGE BOOLE. Boolean algebras were not only of interest for the logicians but also for the set theorists, and later on they found important applications in the theory of automata and also in electrical engineering. Parallel to this development the interest in non-classical logic increased, and also for these logics the algebraic counterparts were defined and investigated. The important property of the algebra of all truth values in such a logic is that its algebraic structure is rich enough to express every operation on the algebra by the algebraic polynomials. Of course one needs the language of universal algebra to give this a precise meaning - not till 1950 was the language of universal algebra developed far enough to define algebras with this property. FOSTER was the first one to investigate these algebras systematically and he called them <u>functionally complete algebras</u>.

In recent years the interest in these algebraic structures has increased rapidly. One reason for this is probably the interest Computer scientists take in these questions, particularly as Computer Science is such a fast growing field of research.

Originally this work was planned as a survey on functionally complete algebras, but then in the last year the theory of functionally complete algebras was extended so much further that the author decided to cover this new development as he had an important part in this. Besides, A.F.PIXLEY, who is one of the leading mathematicians in this field, is about to write a survey-article about functional completeness.

In the theory of functionally complete algebras the ternary discriminator - that is the ternary operation defined by $t(x,y,z) = \begin{cases} x & \text{if } x \neq y \\ z & \text{if } x = y \end{cases}$ - plays a fundamental rôle (see [170]). We will see in the sequel that the ternary discriminator is very powerful and the whole work is dedicated to the investigation of this particular operation.

In the first chapter we give the basic definitions for functionally complete algebras and their special cases. We also give the various characterization theorems due to FOSTER, MAURER-RHODES, PIXLEY, POST, QUACKENBUSH, SHEFFER, SIERPINSKI, WEBB, WILLE, and the author. We conclude the first chapter with a large list of examples. In the second section we investigate the algebraic properties of discriminator varieties, that is varieties of algebras which are generated by algebras as considered in §1. Here we use and extend results in QUACKENBUSH [127]. We then investigate to some extent the varieties generated by the examples given at the end of §1.

The third chapter provides the technical details for a sheaf representation of the algebras in a discriminator variety which we develop in §4. Most of the techniques were given by COMER in [32] and they are also important for the last section.

In the fourth section we extend the sheaf representation from KEIMEL-WERNER [88] to the more general case of discriminator varieties. This sheaf representation is the main tool to investigate the model theoretic properties of the discriminator varieties in the last two sections. In §5 we study the special objects in these varieties such as injective, algebraically closed, existentially closed, or equationally compact algebras. These investigations then lead to some results about the existence of model-companions and model-completions for the first-order theory of such a variety. In the last section we turn to the question of decidable theories. We prove that each residually small discriminator variety has a decidable theory, which is a common generalization of practically all known decidability results for varieties.

Acknowledgements

I am greatly indebted to Dr.Stanley Burris and the Department of Pure Mathematics, University of Waterloo, for their invitation in January - March 1975 , because only the lively discussion and the great interest in their seminar made this work possible. Many thanks I owe to the Fachbereich Mathematik,Technische Hochschule Darmstadt and the Deutsche Forschungsgemeinschaft who made it possible for me to accept the invitation.

§1 FUNCTIONALLY COMPLETE ALGEBRAS

Throughout this work we are concerned with quasi-primal algebras and the varieties they generate. Quasi-primal algebras are special functionally complete algebras, so we start this paper with a chapter about the fundamental properties of functionally complete algebras. In order to keep this paper as self-contained as possible we give some proofs which can also be found elsewhere in the literature.

The notion of functional completeness was introduced by FOSTER [45-54] in 1953; however special functionally complete algebras were already studied much earlier by SHEFFER [149] (1913), POST [124] (1920), ŁUKASIEWICZ [99] (1921), and WEBB [166,167] (1935).

The term changed its meaning later on. In Foster's older terminology, which was influenced by eastern european schools, the expressions *functional completeness* and *weak functional completeness* are used, which in his later works and in the papers of his successors changed to *primality* and *functional completeness* respectively. So the reader should be aware that in papers before 1955,and in more current russian papers,the term "functionally complete" means the same as "primal" in this work.

The problems leading to the investigation of functionally complete algebras come from different parts of mathematics. The first approach was the study of the n-valued propositional logic by POST [124] . Here the Post-algebras play the rôle of the Boolean algebras in the classical propositional logic. The n-element chain considered as a Post-algebra shares the following property with the two-element Boolean algebra: each operation (on the n-element chain) can be expressed by a polynomial (=term) in the language of Post-algebras. An algebra with this property will be called primal.

The second approach is the question of which operations on a given algebra are polynomials. This particular aspect was mostly investigated by the eastern european schools (see ROSENBERG [131 -145], ROUSSEAU [146], PÖSCHEL [121,122], JABLONSKIĬ [70-72], and KALUZNIN et al. [6-8,15,77-81,95,154]).

One way to attack this problem has been to approximate the right answer in the following way. Pick a set of conditions all polynomials satisfy and consider the clone of operations satisfying these conditions, then characterize those algebras for which this is already the clone of all polynomials. This approach has been taken by FOSTER and PIXLEY [45-57,113-120], and their successors [9,27,64-69,88,90-93,124-128,151,152,170-177]. These investigations lead to many weaker notions of primality, almost all of which are stronger than functional completeness. As we shall see, for all these notions the ternary discriminator plays a crucial rôle.

The third approach comes from computer science and the theory of automata, where one of the basic questions is: which operations can be computed on a given machine? In [179] WILLE gave an algebraic version of this problem by interpreting a machine as an algebra, so the above question turns into the same algebraic question as in the second approach.

In this section we give the basic definitions and properties of primal algebras and their generalizations. We also give the characterizations of these special algebras, most of which can be found in the literature with various different proofs. We will give here a uniform proof and thereby get a new characterization of demi-primal algebras.

The basic idea for the whole investigation in this field is that polynomials can be characterized by their "preservation properties" in the sense of proposition 1.1 below.
For any set X, $n\in\mathbb{N}$ and $D\subseteq X^n$ we call D an <u>*n-ary relation*</u> on X and each mapping $f: D\longrightarrow X$ we call a <u>*partial n-ary operation*</u> on X with <u>*domain*</u> $domf := D$ if $domf = X^n$ we call f just an <u>*n-ary operation*</u> on X.
$f: X^n \longrightarrow X$ is said to <u>*preserve*</u>

(i) an m-ary relation $R\subseteq X^m$ iff

$$(x_1^1,\dots,x_1^m),\dots,(x_n^1,\dots,x_n^m)\in R \text{ implies}$$
$$(f(x_1^1,\dots,x_n^1),\dots,f(x_1^m,\dots,x_n^m))\in R .$$

(ii) a partial m-ary operation g on X iff

$(x_1^1,\ldots,x_1^m),\ldots,(x_n^1,\ldots,x_n^m)\in dom g$ implies

a) $(f(x_1^1,\ldots,x_n^1),\ldots,f(x_1^m,\ldots,x_n^m))\in dom g$ and

b) $f(g(x_1^1,\ldots,x_1^m),\ldots,g(x_n^1,\ldots,x_n^m))=$
$g(f(x_1^1,\ldots,x_n^1),\ldots,f(x_1^m,\ldots,x_n^m))$.

Note that f preserves the partial operation g iff f preserves the m+1-ary relation
$\{(x_1,\ldots,x_m,g(x_1,\ldots,x_m))\mid (x_1,\ldots,x_m)\}\in dom g$.
We also permit m to be infinite.

A universal algebra $\underline{A}$ is a pair (A,F) consisting of a set A, which is called the *underlying set* of $\underline{A}$, and a family F of operations on A , which are called the *fundamental operations* on $\underline{A}$. The n-ary *polynomials* are those n-ary operations on A which can be obtained by finitely many compositions from the fundamental operations and the *trivial operations* $e_i^n(x_1,\ldots,x_n)=x_i$ (i=1,..,n) on A . The n-ary *algebraic functions* on $\underline{A}$ are those n-ary operations on A which can be obtained by finitely many compositions from the fundamental operations, the trivial operations, and the *constant operations* $c_a^n(x_1,\ldots,x_n)=a$ $(a\in A)$. In other words the n-ary polynomials on $\underline{A}$ form the subalgebra $\underline{P}^n(\underline{A})$ of $\underline{A}^{A^n}$ which is generated by $\{e_1^n,\ldots,e_n^n\}$, and likewise the n-ary algebraic functions on $\underline{A}$ form the subalgebra $\underline{F}^n(\underline{A})$ of $\underline{A}^{A^n}$ which is generated by $\{e_1^n,\ldots,e_n^n\}\cup\{c_a^n\mid a\in A\}$.

It is quite obvious that each polynomial on $\underline{A}$ preserves all subalgebras of each power $\underline{A}^I$ of $\underline{A}$, and the following proposition shows that also the converse is true.

1.1 PROPOSITION: Let $\underline{A}$ be an algebra and $f:A^n\longrightarrow A$ an operation on A.
f is a polynomial on $\underline{A}$ iff f preserves all subalgebras of $\underline{A}^{A^n}$.

Proof: We see in fact that f has only to preserve the one subalgebra $\underline{P}^n(\underline{A})$ of $\underline{A}^{A^n}$, namely $f=f(e_1^n,\ldots,e_n^n)\in\underline{P}^n(\underline{A})$ if f preserves $\underline{P}^n(\underline{A})$. □

We now know that we can recognize the polynomials on an algebra $\underline{A}$ by their presevation properties on high enough powers of $\underline{A}$, and the question arises, whether one can bound the power to which one has to go. A most beautiful theorem of BAKER-PIXLEY [9] says that in many important cases the polynomials on an algebra are already determined by their preservation properties on the square $\underline{A}\times\underline{A}$ of the algebra $\underline{A}$.

1.2 THEOREM (BAKER-PIXLEY [9]):

Let $\underline{A}$ be a finite algebra with a ternary polynomial d satisfying the equations $d(x,x,y)=x=d(x,y,x)=d(y,x,x)$. Then each operation f on A, which preserves all subalgebras of $\underline{A}\times\underline{A}$, is a polynomial on $\underline{A}$.

Proof: Let $f:A^n \longrightarrow A$ be an operation on A, which preserves all subalgebras of $\underline{A}\times\underline{A}$. We denote by $\mathcal{J}$ the set of all subsets $D\subseteq A^n$ such that there is a polynomial p on $\underline{A}$, which *interpolates* f on D, i.e. $x\in D \Rightarrow f(x)=p(x)$, and we want to show $A^n\in\mathcal{J}$. We prove by induction over $k\geq 2$:

$|D|\leq k \;\&\; D\subseteq A^n \Rightarrow D\in\mathcal{J}$.

k=2: $D=\{(x_1,\ldots,x_n),(y_1,\ldots,y_n)\}$, $x:=f(x_1,\ldots,x_n)$, $y:=f(y_1,\ldots,y_n)$.

f preserves all subalgebras of $\underline{A}\times\underline{A}$, so (x,y) is in the subalgebra generated by $\{(x_1,y_1),\ldots,(x_n,y_n)\}$, so there is a polynomial p on $\underline{A}$ such that $x=p(x_1,\ldots,x_n)$ and $y=p(y_1,\ldots,y_n)$, which puts D into $\mathcal{J}$.

k→k+1: Let $|D|=k+1\geq 3$ and pick $x_1,x_2,x_3\in D$ all different. By the induction hypothesis f can be interpolated on $D\setminus\{x_i\}$ by a polynomial p_i $(i=1,2,3)$. It is now easy to check that the polynomial $d(p_1,p_2,p_3)$ interpolates f on D. □

The notion of primality and its generalizations are all defined by means of subalgebras of the square of an algebra. These systems are:

$S(\underline{A}) = \{B\subseteq A \mid B \text{ subalgebra of } \underline{A}\}$

$Aut(\underline{A}) = \{\phi:A\longrightarrow A \mid \phi \text{ automorphism of } \underline{A}\}$

$Iso(\underline{A}) = \{\phi : B \longrightarrow C \mid B, C$ subalgebras of $\underline{A}$, ϕ isomorphism$\}$

$C(\underline{A}) = \{\Theta \subseteq A \times A \mid \Theta$ congruence on $\underline{A}\}$

Let $\underline{A}$ be a finite algebra. If $\underline{A}$ has the property that the n-ary ($n \neq 0$) polynomials on $\underline{A}$ are precisely those n-ary operations on A which preserve all members of:

$\emptyset$	then $\underline{A}$ is called	*primal* (FOSTER [47])
$S(\underline{A})$	then $\underline{A}$ is called	*semi-primal* (FOSTER-PIXLEY [55])
$Aut(\underline{A})$	then $\underline{A}$ is called	*demi-primal* (QUACKENBUSH [126])
$S(\underline{A}) \cup Aut(\underline{A})$	then $\underline{A}$ is called	*demi-semi-primal* (QUACKENBUSH [126])
$Iso(\underline{A})$	then $\underline{A}$ is called	*quasi-primal* (PIXLEY [115]) (*simple-algebraic* PIXLEY [114])
$C(\underline{A})$	then $\underline{A}$ is called	*hemi-primal* (FOSTER [53])

FOSTER [52,54] also introduced two different notions of infra-primal algebras which do not completely fit into this framework. They were introduced as the natural generalization of finite fields. A finite algebra $\underline{A}$ is called (*strongly*) *infra-primal* iff it is demi-semi-primal and each automorphism of $\underline{A}$ preserves all subalgebras (and all automorphisms) of $\underline{A}$.
There are two more related notions which deal with algebraic functions rather than polynomials.
A finite algebra $\underline{A}$ is called *functionally complete* iff every operation on A is an algebraic function on $\underline{A}$ (FOSTER [47]).
An arbitrary algebra is called *affine complete* iff every operation on A which preserves all congruences on $\underline{A}$ is an algebraic function on $\underline{A}$ (WERNER [171]).

All these special algebras have been characterized except for the case of hemi-primal and affine complete algebras, where only partial solutions are known even in the finite case. We are here giving direct proofs of the known characterizations. Primal algebras and functionally complete algebras will be characterized by the existence of certain polynomials or algebraic functions derived from a general composition theorem for operations.
Here the ternary discriminator plays an important rôle. We investigate the properties of the discriminator and then give a similar characterization for quasi-primal algebras and thereby also for the special quasi-primal algebras (1.14).

1.3 THE COMPOSITION THEOREM FOR OPERATIONS:

(WILLE [179] , WERNER [170])

Let A be a finite set, and $0,1 \in A$ such that $0 \neq 1$.

Let $\wedge$ and $\vee$ be binary operations on A satisfying $a \wedge 1 = a$, $a \wedge 0 = 0$, $a \vee 0 = a = 0 \vee a$, and for every $a \in A$ let χ_a be a unary operation on A such that $\chi_a(x) = \begin{cases} 1 & \text{if } x=a \\ 0 & \text{if } x \neq a \end{cases}$.

Write $\hat{x}$ for $(x_1,\dots,x_n)$, $\hat{a}$ for $(a_1,\dots,a_n)$, and put $\chi_{\hat{a}}(\hat{x}) := (..(\chi_{a_1}(x_1) \wedge \chi_{a_2}(x_2)) \wedge ..) \wedge \chi_{a_n}(x_n)$.

Then every operation $f: A^n \longrightarrow A$ can be written in the following way : $f(\hat{x}) = \bigvee\{c^n_{f(\hat{a})}(\hat{x}) \wedge \chi_{\hat{a}}(\hat{x}) \mid \hat{a} \in A^n\}$.

Proof: Although $\vee$ is neither commutative nor associative there is no ambiguity in our notation $\bigvee\dots$ because all the terms in the iterated $\vee$ are 0 except possibly one, as $\chi_{\hat{a}}(\hat{x}) = 0$ if $\hat{a} \neq \hat{x}$ and 1 otherwise . □

1.4 COROLLARY (SIERPINSKI [150]): Let $\underline{A}$ be a finite algebra.

If every binary operation is a polynomial on $\underline{A}$ then also all n-ary $(n \geq 1)$ operations are polynomials on $\underline{A}$ (i.e. $\underline{A}$ is primal).

This statement also holds for infinite algebras, but there it is a trivial consequence of the fact that there is a bijection from A^n onto A.

1.5 COROLLARY (FOSTER [47]): Let $\underline{A}$ be a finite algebra with at least two elements.

$\underline{A}$ is primal iff there is a binary polynomial $\cdot$ on $\underline{A}$ such that $(A,\cdot)$ is a *group with zero* and a unary polynomial g which is a cyclic permutation on A.

$(A,\cdot)$ is a group with zero iff $(xy)z = x(yz)$ and $\exists 0 \exists 1\ (x0=0=0x\ \&\ x1=x=1x\ \&\ \forall x \exists y\ (x \neq 0 \Rightarrow xy=1=yx))$.

Proof: Let $|A|=n$. Then $x^{n-1} = \begin{cases} 0 & \text{if } x=0 \\ 1 & \text{if } x \neq 0 \end{cases}$, where 1 is the unit element of the group $A \setminus \{0\}$. Let $k \leq n$ be such that $g^k(0)=1$. Then $0 = x^{n-1} \cdot g^{n-k}(x^{n-1})$, and as g is cyclic we get all the constants as polynomials $g(0)$, $g^2(0)$, ..., $g^n(0)$.

$\chi_r(x) := g^k(g^{n-k}(0) \cdot (g^{n-r}(x))^{n-1}) = \begin{cases} 1 & \text{if } x = g^r(0) \\ 0 & \text{otherwise} \end{cases}$.

Now choose $x \vee y := g^{n-k}(g^k(x) \cdot g^k(y))$ and $x \wedge y := xy$ and apply 1.3 . □

<u>1.6 COROLLARY:</u> Let $\underline{L}$ be a finite lattice with a unique coatom c and let $g: L \longrightarrow L$ be a cyclic permutation with $g(c)=1$ and $g(1)=0$.

Then $(L;\wedge,g)$ is primal.

<u>Proof:</u> $0=g(x)\wedge g^2(x)\wedge\ldots\wedge g^n(x)$ $(n=|L|)$. Thus we have all constant polynomials. Choose $x \vee y := g(g^{n-1}(x)\wedge g^{n-1}(y))$ and $\chi_i(x) := g^{n-1}(g(0)\wedge g^{n-i}(x)) = \begin{cases} 1 & \text{if } x=g^i(0) \\ 0 & \text{otherwise} \end{cases}$ and apply 1.3 . □

<u>1.7 COROLLARY (POST [123]):</u> $\underline{P}_n:=(\{0,\ldots,n-1\};\min,')$, where $x'\equiv x+1 \pmod n$ is a primal algebra.

<u>1.8 COROLLARY (SHEFFER [149] , WEBB [166,167]):</u>

On the set $\{0,\ldots,n-1\}$ one defines the generalized <u>*Sheffer-stroke*</u> : $x|y \equiv \min\{x,y\}+1 \pmod n$.

The algebra $\underline{S}_n := (\{0,\ldots,n-1\};|)$ is primal.

<u>1.9 COROLLARY (WERNER [170]):</u> A finite algebra $\underline{A}$ is functionally complete iff the <u>*ternary discriminator*</u> $t(x,y,z) = \begin{cases} x & \text{if } x\neq y \\ z & \text{if } x=y \end{cases}$ is an algebraic function on $\underline{A}$.

<u>Proof:</u> $|A|=1$ is trivial, so choose $0,1\in A$, $0\neq 1$ and define $x\wedge y := t(y,1,x)$, $x\vee y := t(x,0,y)$, $\chi_0(x) := t(0,x,1)$ and for $a\neq 0$ $\chi_a(x) := t(0,t(a,x,0),1)$; then 1.3 gives the result. □

In this whole paper the ternary discriminator plays the fundamental rôle, so we start with some properties of the ternary discriminator.

<u>1.10 LEMMA:</u> Let $\underline{A}$ be an algebra with the ternary discriminator t as a polynomial. Then the following hold:

(1) $\underline{A}$ and all subalgebras of $\underline{A}$ are simple.

(2) The <u>*normal transform*</u> $n(x,y,u,v) = \begin{cases} u & \text{if } x=y \\ v & \text{if } x\neq y \end{cases}$ (GRÄTZER[60])

and the *switching function* $s(x,y,z) = \begin{cases} x & \text{if } x\neq y,z \\ y & \text{if } x=z \\ z & \text{if } x=y \end{cases}$

are polynomials on $\underline{A}$.

(3) On $\underline{A}$ the following equations hold

(a) $t(x,y,y)=x=t(x,y,x)=t(y,y,x)$

(b) $n(x,y,u,v)=t(t(x,y,u),t(x,\beta,v),v)$

$t(x,y,z) = n(x,y,z,x)$

(c) $s(x,y,z) = n(x,y,z,t(x,z,y))$

$s(x,y,z)=z \iff x=y$

(d) $t(x,y,t(x,y,z)) = t(x,y,z) = t(x,y,t(y,x,z))$

(e) $t(x,y,t(x,z,y)) = x = t(x,y,t(y,z,x))$

(f) $t(x,y,t(z,x,y)) = t(x,y,z) = t(x,y,t(z,y,x))$

(g) $s(x,y,z) = s(x,z,y)$.

(4) There is a translation T which assigns to each open Horn-formula α with free variables $x_1,\dots,x_n$ in the language of $\underline{A}$ an n-ary polynomial $p=T\alpha$ over $\underline{A}$ such that for $a_1,\dots,a_n \in A$ we have

(*) $\underline{A} \models \alpha(a_1,\dots,a_n)$ iff $\underline{A} \models p(a_1,\dots,a_n)=a_1$.

The translation T can be chosen as follows

$T(p=q) := s(p,q,x_1)$

$T(\alpha\vee\beta) := n(T\alpha,x_1,x_1,T\beta)$

$T(\alpha\wedge\beta) := t(T\alpha,x_1,T\beta)$

$T(\alpha\rightarrow\beta) := t(x_1,T\alpha,T\beta)$.

Proof: (1) assume Θ is a congruence on some subalgebra $\underline{B}$ of $\underline{A}$, $(x,y)\in\Theta$, $x\neq y$. Then for each $z\in B$ we have $x=t(x,y,z)\Theta t(x,x,z)=z$, so $\underline{B}$ is simple. (2) is in (3b,c).

(3) (a) is trivial for $x=y$, and for $x\neq y$ (a) immediately follows from the definition. (b)-(g) are easy consequences of (a) for $x=y$, and for $x\neq y$ they can be easily derived from the definitions of t, n, and s.

(4) Use $s(x,y,z)=z \iff x=y$ to start an induction on the complexity of α to prove (*). □

We want to give A.F.Pixley's characterization of quasi-primal algebras and its consequences for the other generalizations of primal algebras. To do this we need the notion of an *arithmetical* algebra, i.e. an algebra, which has permutable and distributive congruences. A variety of algebras is called *arithmetical* if all its members are arithmetical. It turns out that it is a Mal'cev-condition (see NEUMANN [109], PIXLEY [116], TAYLOR [164], or

WILLE [178] for this notion) for a variety to be arithmetical - namely the following holds :

1.11 THEOREM (PIXLEY [113]): For a variety $\mathfrak{A}$ t.f.a.e.

(1) $\mathfrak{A}$ is arithmetical.

(2) The free algebra $\underline{F}(3,\mathfrak{A})$ on 3 generators in $\mathfrak{A}$ is arithmetical.

(3) There exists a ternary polynomial t on $\mathfrak{A}$ satisfying $t(x,y,y) = x = t(x,y,x) = t(y,y,x)$.

(4) There exist ternary polynomials p,d on $\mathfrak{A}$ satisfying $p(x,y,y)=x=p(y,y,x)$ & $d(x,x,y)=x=d(x,y,x)=d(y,x,x)$.

In arithmetical varieties we have in particular permutable congruences and a result of I.Fleischer says that in such classes subdirect products are of a very special kind, roughly speaking a subdirect product of $\underline{A}$ and $\underline{B}$ is (the graph of) an isomorphism between quotients of $\underline{A}$ and $\underline{B}$.

1.12 THEOREM: For a variety $\mathfrak{A}$ of algebras t.f.a.e.

(1) Each $\underline{A}\in\mathfrak{A}$ has permutable congruences.

(2) (MAL'CEV [103]) There is a ternary polynomial p on $\mathfrak{A}$ satisfying $p(x,y,y) = x = p(y,y,x)$.

(3) (FLEISCHER [44]) If $\underline{R}$ is a subalgebra of $\underline{A}\times\underline{A}$ then there exist subalgebras $\underline{B},\underline{C}$ of $\underline{A}$, cougruences $\Theta\in C(\underline{B})$, $\Phi\in C(\underline{C})$, and an isomorphism $f:\underline{B}/\Theta \to \underline{C}/\Phi$ such that $R = \{(b,c) \mid b\in B, c\in C, f([b]\Theta) = [c]\Phi\}$.

(4) (WERNER [172]) Each reflexive subalgebra of $\underline{A}\times\underline{A}$ is a congruence on $\underline{A}\in\mathfrak{A}$.

(This result was found independently by Shafaat in Can.Math, Bull. 17(1974),609)

In the sequel we need a characterization of the pairs $(Aut(\underline{A}),S(\underline{A}))$ for all finite algebras $\underline{A}$, which is a slight modification of the theorem of STONE [155].

1.13 THEOREM (STONE [155]): Let A be a finite set, G a permutation group on A, and H a closure system on A. The following conditions are equivalent:

(1) There is an algebraic structure $\underline{A}$ on A such that $G = Aut(\underline{A})$ and $H = S(\underline{A})$

(2) a) $M \in H$ & $g \in G \Rightarrow gM := \{gx \mid x \in M\} \in H$

b) $g \in G \Rightarrow F := \{x \in A \mid gx = x\} \in H$

(3) There is an algebraic structure $\underline{A}$ on A such that $Iso(\underline{A}) = \{h \mid domh \in H \ \& \ (|domh| = 1 \text{ or } \exists g \in G \ h = g|domh)\}$

Proof: (3)=>(1)=>(2) are trivial, so we have to show (2)=>(3). Observe that under the assumptions in (2), if a is in the H-closure of $a_1, \ldots, a_n$ then the definition

$f(x_1, \ldots, x_n) := \begin{cases} g(a) & \text{if } g \in G \ \& \ g(a_1) = x_1, \ldots, g(a_n) = {}_n \\ x_1 & \text{otherwise} \end{cases}$ defines an

n-ary operation on A by (2b), which by definition preserves each $g \in G$, and by (2a) preserves each $M \in H$.

Let $\underline{A}$ be the algebra on A with all operations of this kind. Obviously $S(\underline{A}) = H$. Now let $M, N \in H$, $h: M \to N$ an isomorphism between the subalgebras M and N. Let $M = \{a_1, \ldots, a_n\}$, $a_1 \neq a_2$. As a_2 is in the closure of M we have a polynomial f on $\underline{A}$ such that

$$f(x_1, \ldots, x_n) = \begin{cases} x_2 & \text{if } \exists g \in G \ (x_1, \ldots, x_n) = (ga_1, \ldots, ga_n) \\ x_1 & \text{otherwise} \end{cases}$$

Now $f(a_1, \ldots, a_n) = a_2$, so $f(ha_1, \ldots, ha_n) = ha_2 \neq ha_1$, thus $\exists g \in G \ (ha_1, \ldots, ha_n) = (ga_1, \ldots, ga_n)$, but this says that h is the restriction $g|M$ of $g \in G$ to M. □

We should remark that one could give an infinite version of the above theorem by demanding in (2) that H is an algebraic closure system and that G is locally closed, i.e. G satisfies the condition in JÓNSSON [74].

In order to have shorter notation we call the sets of the form $Fg_1 \cap \ldots \cap Fg_n$ in (2b) *fixpoint-sets of G*, and the isomorphisms between subalgebras of $\underline{A}$ we call *inner isomorphisms* of $\underline{A}$. We say an inner isomorphism is *proper* if its domain has at least two elements and it is not the identity map on its domain.

We are now ready to state Pixley's characterization of quasi-primal algebras and as a consequence also the characterizations of more special quasi-primal algebras due to various authors.

1.14 THEOREM: Let $\underline{A}$ be a finite algebra and $\mathfrak{a}$ the variety generated by $\underline{A}$.

(1) (PIXLEY [115]) The following are equivalent

(i) $\underline{A}$ is quasi-primal.

(ii) The ternary discriminator is a polynomial on $\underline{A}$.

(iii) $\mathfrak{a}$ is arithmetical and all subalgebras of $\underline{A}$ are simple.

(2) (QUACKENBUSH [128]) The following are equivalent:

(i) $\underline{A}$ is demi-semi-primal.

(ii) $\mathfrak{a}$ is arithmetical, all subalgebras of $\underline{A}$ are simple, and each proper inner isomorphism of $\underline{A}$ extends to an automorphism of $\underline{A}$.

(3) (FOSTER [52,54]) The following are equivalent:

(i) $\underline{A}$ is (strongly) infra-primal.

(ii) $\mathfrak{a}$ is arithmetical, all subalgebras of $\underline{A}$ are simple, no two different subalgebras of $\underline{A}$ are isomorphic, and each automorphism of a subalgebra of $\underline{A}$ extends to an automorphism of $\underline{A}$ (and $Aut(\underline{A})$ is an abelian group).

(4) (FOSTER-PIXLEY [55]) The following are equivalent:

(i) $\underline{A}$ is semi-primal .

(ii) $\mathfrak{a}$ is arithmetical, all subalgebras of $\underline{A}$ are simple, no two different subalgebras of $\underline{A}$ are isomorphic, and no subalgebra of $\underline{A}$ has a proper automorphism.

(5) The following are equivalent:

(i) $\underline{A}$ is demi-primal .

(ii) $\mathfrak{a}$ is arithmetical, all subalgebras of $\underline{A}$ are simple, each proper inner isomorphism of $\underline{A}$ extends to an automorphism of $\underline{A}$, and the only subalgebras of $\underline{A}$ are the fixpoint-sets of $Aut(\underline{A})$.

(6) (FOSTER-PIXLEY [55]) The following are equivalent:

(i) $\underline{A}$ is primal .

(ii) $\mathfrak{a}$ is arithmetical, $\underline{A}$ is simple, and $\underline{A}$ has no proper subalgebras and no proper automorphisms .

(7) (PIXLEY [119]) The following are equivalent for a finite arithmetical algebra $\underline{A}$:

(i) $\underline{A}$ is hemi-primal .

(ii) $\mathfrak{A}$ is arithmetical, $\underline{A}$ has no proper subalgebras, and $\underline{A}$ has no proper automorphism.

For the special case that $C(\underline{A})$ is a chain this is due to FOSTER [53].

<u>Proof</u>: (1):(i)=>(ii) is clear as t preserves all subsets and all bijections between them. (ii)=>(iii): By 1.10(3a) and 1.11 $\mathfrak{A}$ is arithmetical and by 1.10(1) all subalgebras of $\underline{A}$ are simple. (iii)=>(i): $\underline{A}\times\underline{A}$ has permutable congruences, so by Fleischer's result 1.12(4) the subalgebras of $\underline{A}\times\underline{A}$ are products of subalgebras of $\underline{A}$ or inner isomorphisms of $\underline{A}$. Each product of subalgebras can be obtained as the relational product of $\underline{A}\times\underline{A}$ and the identities on the subalgebras, so the operations which preserve all subalgebras of $\underline{A}\times\underline{A}$ are precisely those, which preserve all inner isomorphisms of $\underline{A}$. Now we can apply the theorem of Baker-Pixley 1.2 to see (iii)=>(i).
(2):(i)=>(ii) follows from 1.13 and (ii)=>(i) from (1).
(3) is the special case of (2), where all automorphisms preserve all subalgebras. Note that automorphisms preserve each other iff they commute.
(4) is the special case of (2) with $Aut(\underline{A}) = id_A$.
(5) is the special case of (2) with $S(\underline{A}) = \{Fg_1 \cap \ldots \cap Fg_n \mid g_i \in Aut(\underline{A})\}$ by 1.13.
(6) is the common specialization of (4) and (5).
(7):(i)=>(ii) By PIXLEY [119] there is a congruence preserving ternary operation t on $\underline{A}$ such that $t(x,y,y)=x=t(x,y,x)=t(y,y,x)$, thus t is a polynomial and $\mathfrak{a}$ is arithmetical by 1.11. All constants are polynomials on $\underline{A}$, so $\underline{A}$ can't have proper subalgebras or proper automorphisms. (ii)=>(i) Use the theorems of Fleischer and Baker-Pixley as in the proof of (1). □

We will see in the sequel that varieties generated by quasi-primal algebras are very special (§2) and generalizing these properties CLARK-KRAUSS [27] were led to the notion of a <u>*para-primal*</u> algebra, which can be characterized by the property that it generates a permutable variety and all its subalgebras are simple. Groups of prime order are typical examples, but we do not go into further details about para-primal algebras.

We now give some short remarks about functionally complete and affine complete algebras. We need the notion of a <u>*skew congruence*</u> on $\underline{A}\times\underline{B}$, that is a congruence which is not a product $\Theta\times\Phi := \{((a,b),(\hat{a},\hat{b})) \mid (a,\hat{a})\in\Theta \ \&\ (b,\hat{b})\in\Phi\}$ of congruences $\Theta\in C(\underline{A})$ and $\Phi\in C(\underline{B})$.

1.15 THEOREM (WERNER [175]): Let $\underline{A}$ be a finite, simple algebra in a congruence-permutable variety.
$\underline{A}$ is functionally complete iff $\underline{A}\times\underline{A}$ has no skew congruence.

1.16 COROLLARY (MAURER-RHODES [104])
A group is functionally complete iff it is a finite, simple, non-abelian group.

Proof of 1.15 : Let $\underline{A}^*$ denote the algebra $\underline{A}$ with all constants added and let $\mathfrak{A}^*$ be the variety generated by $\underline{A}^*$ We have to show that $\underline{A}^*$ is primal or equivalently that $\mathfrak{A}^*$ is arithmetical and that $\underline{A}^*$ is simple, has no proper subalgebras and no proper automorphisms. All these conditions are trivial from the assumptions except the congruence-distributivity of $\mathfrak{A}^*$. By 1.11(2) it suffices to show that $\underline{F}(3,\mathfrak{A}^*)$ has distributive congruences.

Claim: $\underline{F} := \underline{F}(3,\mathfrak{A}^*)$ is a finite power of $\underline{A}^*$.

$\underline{F}$ is a subdirect power of finitely many copies of $\underline{A}^*$, as $\underline{A}^*$ is finite and has no subalgebra. Assume we have an irredundant subdirect representation of $\underline{F}$: $p_i:\underline{F} \twoheadrightarrow \underline{A}^*$ $1\le i\le n$, meaning that for each $i=1,..,n$ the p_j $(j\ne i)$ no longer separate F. As $\underline{A}^*$ is simple, the kernels of the p_i's are maximal congruences, they permute with each other and they are easily seen to satisfy the conditions in BIRKHOFF [14,VII §4] for a direct decomposition, in other words $\underline{F} \cong \underline{A}^{*n}$.

We now use the result of WERNER [175] that, in a congruence permutable variety, an algebra whose square has no skew congruence has no skew congruence in any finite power, so in our case $C(\underline{F}) \cong C(\underline{A}^{*n}) \cong [C(\underline{A}^*)]^n = \mathbf{2}^n$ is distributive . □

For affine complete algebras we have a corollary of the result of Baker-Pixley and an extension of the result of HU [65] on Boolean powers (see 5.14).

1.17 THEOREM (BAKER-PIXLEY [9]): In an arithmetical variety $\mathfrak{A}$ each finite algebra is affine complete.

Proof: By Baker-Pixley 1.2 the algebraic functions on a finite $\underline{A} \in \mathfrak{A}$ are precisely those operations which preserve all reflexive subalgebras of $\underline{A} \times \underline{A}$, which are the congruences on $\underline{A}$ by Werner 1.12(5). □

1.18 THEOREM: If $\underline{A}$ is functionally complete and B is a Boolean algebra, then the Boolean power $\underline{A}[B]$ is affine complete.

Proof: Consider $\underline{A}^*$ as in the proof of 1.15. $\underline{A}^*$ is primal, so by Hu's result 5.14 $\underline{A}^*[B]$ is affine complete. But $\underline{A}[B]$ and $\underline{A}^*[B]$ have the same non-constant operations. □

In the sequel we need the notion of independent algebras due to FOSTER [46]. Algebras $\underline{A}_1,\dots,\underline{A}_n$ are called *independent* if for every set of k-ary polynomials $p_1,\dots,p_n$ there is a polynomial p such that for $i=1,\dots,n$

$\underline{A}_i \models \forall x_1 \dots \forall x_k \; p_i(x_1,\dots,x_k) = p(x_1,\dots,x_k)$

1.19 PROPOSITION: For quasi-primal algebras $\underline{A}_1,\dots,\underline{A}_n$ t.f.a.e.

(1) $\underline{A}_1,\dots,\underline{A}_n$ are independent.

(2) There is an n-ary polynomial e such that for $i=1,\dots,n$ $\underline{A}_i \models \forall x_1,\dots,x_n \; e(x_1,\dots,x_n) = x_i$.

(3) a) There is a ternary polynomial t which is the discriminator on each of $\underline{A}_1,\dots,\underline{A}_n$.

b) If $i \neq j$ then $\underline{A}_i$ and $\underline{A}_j$ have no isomorphic non-singleton subalgebras.

Proof: (1)<=>(2) holds in general for every finite set $\underline{A}_1,\dots,\underline{A}_n$ of algebras. (1)=>(2) is trivial and for (2)=>(1) choose $p = e(p_1,\dots,p_n)$. (1)=>(3a): choose the polynomial t corresponding to the discriminators t_1 on $\underline{A}_1$, ..., t_n on $\underline{A}_n$. (1)=>(3b) if $i \neq j$ and $\underline{A}_i$ and $\underline{A}_j$ had a non-singleton subalgebra in common, then choosing $p_i(x,y) = x$ and $p_k(x,y) = y$ for $k \neq i$, they can't be represented by one single polynomial because on that particular subalgebra it must be the first and second projection at the same time, which is absurd.

(3)=>(2) Turn the disjoint union of $\underline{A}_1,\dots,\underline{A}_n$ into an algebra $\underline{A}$ by defining all fundamental operations $f(x_1,\dots,x_k)$ as on $\underline{A}_i$ if $x_1,\dots,x_k \in A_i$, and x_1 otherwise, and define a new discriminator t^* on $\underline{A}$. Then $\underline{A}$ is quasi-primal and its subalgebras are precisely the disjoint unions of subalgebras of some of the $\underline{A}_i$'s . The operation $f(x_1,\dots,x_n) = x_i$ iff $x_1,\dots,x_n \in A_i$ and $f(x_1,\dots,x_n) = x_1$ otherwise obviously preserves all subalgebras of $\underline{A}$ and all isomorphisms between them, so f is a polynomial over $\underline{A}$. Now replace in f each occurence of t^* by t and you obtain the desired polynonial e in (2) . □

In [57] Foster and Pixley introduced the notion of *weak independence* , which for quasi-primal algebras boils down to the condition (3a) (see also 2.4)

Before we develop our theory any further we want to give some examples of algebras which have the discriminator as a polynomial.

1.20 EXAMPLES :

(1) Let $(H;\wedge,\vee,\rightarrow,0,1)$ be a Heyting-algebra with an additional unary operation $d:H\to H$ such that $d(x) = \begin{cases} 1 & \text{if } x=1 \\ 0 & \text{if } x\neq 1 \end{cases}$. Then the following polynomial is the ternary discriminator on H:

$$t(x,y,z) = [z\wedge d(x\vee z \rightarrow x\wedge z)]\vee[x\wedge(d(x\vee z \rightarrow x\wedge z)\rightarrow 0)] .$$

(2) The 2-element Boolean algebra $\underline{2} = (\{0,1\};\wedge,\vee,')$ has the above discriminator with $d(x)=x$, but there is aso a simpler one:

$t(x,y,z) = ((x\wedge z)\vee y')\wedge(x\vee z)$

(3) The 2-element relatively complemented distributive lattice $\underline{R}_2 = (\{0,1\};\wedge,\vee,r)$, where $r(x,y,z)$ is the relative complement of x in the interval $[y\wedge z, y\vee z]$, has the discriminator (by example 2):

$t(x,y,z) = r(y,x,z)$

(4) The n-element Łukasiewicz-algebra $\underline{L}_n = (\{0,\dots,n-1\};\wedge,\vee,{}^-,D_1,\dots,D_{n-1},0,n-1)$ with the operations $\bar{x} = n-1-x$ and $D_i(x) = \begin{cases} n-1 & \text{if } i\le x \\ 0 & \text{if } i>x \end{cases}$

is a special case of (1) because

$x \to y = y \vee \bigwedge\{\overline{D_i(x)} \vee D_i(y) \mid 1 \le i \le n-1\}$ and $d(x) = D_{n-1}(x)$.

(5) The n-element Post-algebra

$\underline{P}_n = (\{0,\dots,n-1\};\wedge,\vee,C,D_1,\dots,D_{n-1},0,\dots,n-1)$ with the operations like above, only $C(x) = \begin{cases} n-1 & \text{if } x=0 \\ 0 & \text{if } x \ne 0 \end{cases}$, is (even without the constants) a special case of (1) as

$x \to y = y \vee \bigwedge\{C(D_i(x)) \vee D_i(y) \mid 1 \le i \le n-1\}$ and $d(x) = D_{n-1}(x)$.

(6) A Heyting algebra $(H;\wedge,\vee,\to,0,1,^+)$ with a $\vee$-irreducible 1 and with a dual pseudocomplementation $^+$ is a special case of (1) with $d(x) = x^{++}$ (because 1 is $\vee$-irreducible it follows $x^+ = \begin{cases} 1 & \text{if } x \ne 1 \\ 0 & \text{if } x=1 \end{cases}$).

(7) The simple cylindric algebras of dimension n $(B;\wedge,\vee,',0,1,C_1,\dots,C_n)$ are characterized by $C_1C_2\dots C_n(x) = \begin{cases} 1 & \text{if } x=1 \\ 0 & \text{if } x \ne 1 \end{cases}$, so they are special cases of (1) .

(8) The simple monadic algebras $(B;\wedge,\vee,',0,;,C)$ are the case $n=1$ of (7).

(9) The simple relation algebras $(R;\wedge,\vee,',0,1,\circ,\breve{},1')$ are characterized by $1 \circ x \circ 1 = \begin{cases} 1 & \text{if } x \ne 0 \\ 0 & \text{if } x=0 \end{cases}$, thus they are special cases of (1) with $d(x) = (1 \circ x \circ 1)'$

(10) Let $(R;+,-,\cdot,0,1,g)$ be a ring with an additional unary operation $g(x) = \begin{cases} 1 & \text{if } x \ne 0 \\ 0 & \text{if } x=0 \end{cases}$. Then the following polynomial is the discriminator

$t(x,y,z) = z + (x - z)g(y - x)$

Special cases of this example are the following examples (11) - (14) .

(11) The finite field $GF(q)$ with $n-1 \equiv 0 \pmod{q-1}$ satisfies $x^{n-1} = \begin{cases} 1 & \text{if } x \ne 0 \\ 0 & \text{if } x=0 \end{cases}$, so choose $g(x) = x^{n-1}$.

(12) The simple biregular rings $(R;+,-,\cdot,0,^+)$, where $^+$ assigns to each element $a \in R$ the central idempotent a^+, which generates the same ideal as a , are unitary rings and satisfy $x^+ = \begin{cases} 1 & \text{if } x \ne 0 \\ 0 & \text{if } x=0 \end{cases}$, so choose $g(x) = x^+$.

(13) The simple Baer*-rings $(R;+,-,\cdot,0,^*)$, where * assigns to each element $a\in R$ the central idempotent a^* , which generates the annihilator of a , are unitary rings and satisfy $x^* = \begin{cases}1 & \text{if } x=0\\ 0 & \text{if } x\neq 0\end{cases}$, so choose $g(x) = x^{**}$.

(14) Consider a skewfield F as a ring with an additional inversion x^{-1} , with $0^{-1} = 0$ say. Then it is a special case of (10) by the choice $g(x) = x\cdot x^{-1}$.

(15) The n-element complementary semigroup $\underline{S}_n = (\{0,\ldots,n-1\};\cdot,*)$ with the operations $x\cdot y = \min\{n-1,x+y\}$ and $x*y=y-\min\{x,y\}$ has the following polynomial as discriminator
$t(x,y,z) = ((x\nabla y)^k * z)\vee([(x\nabla y)^k * (y\nabla z)^k] * x)$
where $k\geq n-1$, $x\nabla y := (x*y)\cdot(y*x)$ and $x\vee y := x\cdot(x*y)$.

For exact definitions and more detailed information the reader is refered to the second section or to [10] for the examples (1) - (6), to [62,107] for the examples (7),(8) , to [26,163] for (9) , to [33] for (12) , to [13] for (13) , and to [16] for example (15).

§2 DISCRIMINATOR VARIETIES

In this chapter we investigate some properties of varieties which are generated by algebras which have the discriminator as a polynomial. These varieties turn out to be of a very special kind but on the other hand a list of examples at the end of this chapter shows that they include many well known varieties of algebras.

2.1 DEFINITIONS AND NOTATIONS:

Let $\mathcal{W}$ be a variety of algebras and $\underline{R}\in\mathcal{W}$. $C(\underline{R})$ denotes the congruence lattice of $\underline{R}$ with the greatest element $\nabla = \underline{R}\times\underline{R}$ and the smallest element $\Delta = \{(x,x)\mid x\in R\}$. For $x,y\in R$ $\Theta(x,y)$ denotes the smallest congruence on $\underline{R}$ which contains (x,y) . The congruences $\Theta(x,y)$ $(x,y\in R)$ are called *principal congruences* . A congruence $\Theta\in C(\underline{R})$ is *compact* if one of the following two equivalent conditions is satisfied:

(a) $\Theta\subseteq\bigvee\{\Phi_i \mid i\in I\} \Rightarrow \exists n\exists i_1\ldots\exists i_n\ \Theta\subseteq\Phi_{i_1}\vee\ldots\vee\Phi_{i_n}$.

(b) Θ is the join of finitely many principal congruences .

A congruence $\Theta\in C(\underline{R})$ is called a *factor congruence* iff there exists a $\Phi\in C(\underline{R})$ such that $\Theta\wedge\Phi = \Delta$ and $\Theta\circ\Phi = \nabla$, and by BIRKHOFF [14, VII, §4] this is equivalent to saying that the natural map $\underline{R}\to\underline{R}/\Theta\times\underline{R}/\Phi$ is an isomorphism.
A congruence $\Theta\in C(\underline{R})$ is called *semisimple* if it is the intersection of maximal congruences of $\underline{R}$. $\underline{R}$ is called *semisimple* iff it is a subdirect product of simple algebras, and that is the case iff Δ is a semisimple congruence on $\underline{R}$.
$\underline{R}$ has *regular* congruences iff $\forall\Theta,\Phi\in C(\underline{R})\forall x\in R\ [x]\Theta=[x]\Phi \Rightarrow \Theta=\Phi$
$\underline{R}$ has *normal* congruences iff $\forall\Theta\in C(\underline{R})\forall x,y\in R\ |[x]\Theta|=|[y]\Theta|$
$\underline{R}$ has *filtral* congruences iff for each subdirect representation $f:\underline{R}\rightarrowtail\prod_{i\in I}\underline{A}_i$ by subdirectly irreducible algebras $\underline{A}_i$ $(i\in I)$ each congruence on $\underline{R}$ is the restriction of a *filter-congruence* on $\Pi\underline{A}_i$, i.e. for $\Theta\in C(\underline{R})$ there exists a filter F on I such that $(x,y)\in\Theta \Leftrightarrow \{i \mid f(x)_i=f(y)_i\}\in F$.
$\mathcal{V}$ is said to have the *congruence extension property* (CEP) iff for every embedding $f:\underline{R}\rightarrowtail\underline{S}$ in $\mathcal{V}$ and $\Theta\in C(\underline{R})$ there is a congruence $\Phi\in C(\underline{S})$ such that
$\Theta = f^{-1}(\Phi) := \{(x,y) \mid x,y\in R\ \&\ (fx,fy)\in\Phi\}$.

Now we are ready to prove the main theorem on *discriminator varieties* which we will encounter later.

2.2 THEOREM : Let $\mathfrak{A}$ be a class of algebras of the same type and let t be a ternary polynomial which is the ternary discriminator on each $\underline{A}\in\mathfrak{A}$.
For the variety $\mathcal{V} = HSP(\mathfrak{A})$ generated by $\mathfrak{A}$ the following hold:

(1) $\mathcal{V}$ is arithmetical .

(2) For $\underline{A}\in\mathcal{V}$, $|A|\geq 2$ the following are equivalent
(i) $\underline{A}$ is subdirectly irreducible .
(ii) $\underline{A}\in SP_U(\mathfrak{A})$.
(iii) t is the discriminator on $\underline{A}$.
(iv) $\underline{A}$ is simple .
(v) $\underline{A}$ is directly indecomposable .

(3) If $\mathfrak{A}$ is finite (up to isomorphism) and every $\underline{A}\in\mathfrak{A}$ is finite then $S\mathfrak{A}$ is (up to isomorphism) the set of all subdirectly irreducible algebras in $\mathfrak{V}$.

(4) Each $\underline{R}\in\mathfrak{V}$ is semisimple and each $\Phi\in C(\underline{R})$ is semisimple.

(5) For $\underline{R}\in\mathfrak{V}$ and $x,y,u,v\in R$ we have
$\Theta(x,y)\subseteq\Theta(u,v) \iff t(u,v,x) = t(u,v,y)$.

(6) Each principal congruence on $\underline{R}\in\mathfrak{V}$ is a factor congruence.

(7) The principal congruences on $\underline{R}\in\mathfrak{V}$ form a sublattice of the congruence lattice $C(\underline{R})$.

(8) Each compact congruence on $\underline{R}\in\mathfrak{V}$ is a principal congruence.

(9) The congruences on each $\underline{R}\in\mathfrak{V}$ are regular.

(10) The congruences on each $\underline{R}\in\mathfrak{V}$ are normal.

(11) The congruences on each $\underline{R}\in\mathfrak{V}$ are filtral.

(12) $\mathfrak{V}$ has the congruence extension property.

Proof: (1): By lemma 1.10 t satisfies the equations $t(x,x,y)=y$ and $t(x,y,x) = x = t(x,y,y)$ and thus by Pixley's theorem 1.11 the variety $\mathfrak{V}$ has permutable and distributive congruences.
(2): (ii)=>(iii): "t is the discriminator on $\underline{A}$" is a first-order universal sentence and thus carries over to all ultraproducts and subalgebras. (iii)=>(iv) is 1.10(1), (iv)=>(i),(v) are trivial and (v)=>(iv) follows from (7) below, so it remains to show (i)=>(ii). $\mathfrak{V}$ has distributive congruences, so by Jõnsson's lemma [73,Cor.3.2] all subdirectly irreducible members of $\mathfrak{V}$ belong to $HSP_U(\mathfrak{A})$, but by (ii)=>(iii)=>(iv) all members of $SP_U(\mathfrak{A})$ are simple and thus every homomorphic image of some $\underline{A}\in SP_U(\mathfrak{A})$ is either isomorphic to $\underline{A}$ or a singleton.
(3): If $\mathfrak{A}$ is a finite set of finite algebras, then $P_U(\mathfrak{A})\subseteq I(\mathfrak{A})$ so (3) follows from (2).
(4) is an immediate consequence of Birkhoff's theorem that every algebra is a subdirect product of subdirectly irreducibles, and (2) that all subdirectly irreducibles are simple.
(5): By (3) the condition $\Theta(x,y) \subseteq \Theta(u,v)$ is equivalent to saying that every maximal congruence Φ, which contains (u,v), also contains (x,y). Now, for every maximal congruence Φ, t is the discriminator on $\underline{R}/\Phi$ and the equation $t(u,v,x) = t(u,v,y)$ is equivalent to saying for every maximal congruence Φ we have

$[u]\Phi=[v]\Phi \Rightarrow [x]\Phi=[y]\Phi$ or in other words $(u,v)\in\Phi \Rightarrow (x,y)\in\Phi$.
(6): Define for $r,s\in R$ $\Psi(r,s) := \{\Phi \mid \Phi$ maximal in $C(\underline{R})$ & $(r,s)\notin\Phi\}$
Then $\Theta(r,s)\cap\Psi(r,s) = \bigcap\{\Phi \mid \Phi$ maximal in $C(\underline{R})\} = \Delta$.
Observe that a maximal congruence Φ satisfies $(r,s)\notin\Phi$ iff for all $x\in R$ we have $(t(r,s,x),r)\in\Phi$, thus $\{(t(r,s,x),r) \mid x\in R\}\subseteq \Psi(r,s)$. This shows that no maximal congruence contains both $\Psi(r,s)$ and $\Theta(r,s)$, hence $\Psi(r,s)\vee\Theta(r,s) = \nabla$. $\Psi(r,s)$ and $\Theta(r,s)$ are complementary to each other and permutable by (1), so they are factor congruences, as $\Psi(r,s)\vee\Theta(r,s) = \Psi(r,s)\circ\Theta(r,s)$
(7): For each maximal congruence Φ we have by (2) that t is the discriminator on $\underline{R}/\Phi$, and thus we have for $x,y,z\in R$
$(x,y)\in\Phi \Leftrightarrow (t(x,y,z),z)\in\Phi$ and $(x,y)\notin\Phi \Leftrightarrow (t(x,y,z),x)\in\Phi$.
From this we get for all $x,y,u,v\in R$:
$(x,y)\in\Phi$ & $(u,v)\in\Phi \Leftrightarrow (t(x,y,u),t(y,x,v))\in\Phi$ and
$(x,y)\in\Phi$ or $(u,v)\in\Phi \Leftrightarrow (t(x,t(x,y,u),u),t(x,t(x,y,v),v)\in\Phi$.
Now we get by (4) $\Theta(x,y)\vee\Theta(u,v) = \Theta(t(x,y,u),t(y,x,v))$ and $\Theta(x,y)\cap\Theta(u,v) = \Theta(t(x,t(x,y,u),u),t(x,t(x,y,v),v))$.
(8) is an immediate consequence of (7).
(9): Let $a\in R$, then we have $(x,y)\in\Phi \Leftrightarrow s(x,y,a)\in[a]\Phi$ for every maximal congruence Φ on $\underline{R}$ by 1.10(3), but then because of (3) the same holds for every congruence on $\underline{R}$.
(10): 1.10(3c) says that the mapping $x\mapsto s(x,a,b)$ is a self-inverse bijection and by (6) maps $[a]\Phi$ into $[b]\Phi$ and vice versa, thus $|[a]\Phi| = |[b]\Phi|$.
(11): Let $f:\underline{R} \rightarrowtail \prod_{i\in I} A_i$ be a subdirect representation of $\underline{R}$ by simple algebras. We first show that all maximal congruences of $\underline{R}$ are filtral, then the result follows because the intersection of a set of filter-congruences is again a filter-congruence. Let Φ be a maximal congruence on $\underline{R}$. Then by Jónsson's lemma [73, lemma 3.1] there is an ultrafilter U on I such that the congruence $\Theta_U := \{(x,y) \mid \{i \mid f(x)_i=f(y)_i\}\in U\}$ is contained in Φ. $\underline{R}/\Theta_U$ is a subalgebra of an ultraproduct of the $\underline{A}_i$'s, hence it is a simple algebra and thus $\Theta_U=\Phi$.
(12). is an immediate consequence of (5) and DAY [37]. (12) is also an immediate consequence of (11). □

This theorem gives rise to the definition of a *discriminator variety*, that is a variety $\mathcal{V}$, which has a ternary polynomial t such that t is the discriminator on every subdirectly irreducible algebra in $\mathcal{V}$. Observe that in a discriminator variety $\mathcal{V}$ the class $\mathcal{A}$ of all subdirectly irreducible algebras in $\mathcal{V}$ satisfies the conditions in 2.2, so a discriminator variety satisfies 2.2(1)-(12) .

2.4 RESIDUAL FINITENESS

A variety $\mathcal{V}$ is called *residually small* if it contains up to isomorphism only a set of subdirectly irreducible algebras, and *residually finite* if all subdirectly irreducible members of $\mathcal{V}$ are finite. In order to characterize the residually small and residually finite discriminator varieties we need the notion of a *weakly independent set of quasi-primal algebras* , that is a finite set $\mathcal{A}$ of finite algebras for which there is a ternary polynomial t which is the discriminator on each $\underline{A} \in \mathcal{A}$.

The following theorem characterizes the residually small discriminator varieties. It can be easily derived from the results in TAYLOR [165], but we give a simple direct proof.

THEOREM: For a discriminator variety $\mathcal{V}$ of finite type t.f.a.e.

(i) $\mathcal{V}$ is generated by a weakly independent set of quasi-primal algebras.

(ii) $\mathcal{V}$ is residually finite.

(iii) $\mathcal{V}$ is residually small.

(iv) $\mathcal{V}$ has up to isomorphism only finitely many simple algebras.

Proof: (i)=>(ii) by 2.2(3) and (ii)=>(iii) is trivial. By 2.2(2) being simple is a first-order property for an algebra in $\mathcal{V}$, so if $\mathcal{V}$ contains an infinite subdirectly irreducible (=simple) algebra or infinitely many finite simple algebras, then each ultraproduct of those is again simple, and we get a proper class of non-isomorphic simple algebras. This proves (iii)=>(iv) and shows in case of (iv) that all the simple algebras have to be finite, but then they form a weakly independent set of quasi-primal algebras and generate $\mathcal{V}$,so this proves (iv)=>(i) . □

2.5 AMALGAMATION PROPERTY

Let $\mathcal{K}$ be a class of algebras of the same type. A pair of embeddings $f:\underline{A} \rightarrowtail \underline{B}$ and $g:\underline{A} \rightarrowtail \underline{C}$ with $A \neq \emptyset$, $\underline{A},\underline{B},\underline{C} \in \mathcal{K}$

is called an *amalgam* in $\mathfrak{K}$, and it is said to be *amalgamated* in $\mathfrak{K}$ if there exists an algebra $\underline{D}$ in $\mathfrak{K}$ and embeddings $h:\underline{B}\rightarrowtail\underline{D}$ and $l:\underline{C}\rightarrowtail\underline{D}$ such that the diagram

$$\begin{array}{ccc} \underline{A} & \rightarrowtail & \underline{B} \\ \downarrow & & \downarrow \\ \underline{C} & \rightarrowtail & \underline{D} \end{array}$$

commutes , i.e. $h\circ f = l\circ g$.

$\mathfrak{K}$ has the *amalgamation property (AP)* , iff every amalgam in $\mathfrak{K}$ can be amalgamated in $\mathfrak{K}$.

The following theorem allows us to reduce the amalgamation problem in a discriminator variety to the simple members. It is due to GRÄTZER-LAKSER [61] , and in this form due to QUACKENBUSH [128] .

(1) THEOREM : Let $\mathfrak{K}$ be a variety of semisimple algebras with the congruence extension property.

The amalgam $f:\underline{A}\rightarrowtail\underline{B}$, $g:\underline{A}\rightarrowtail\underline{C}$ in $\mathfrak{K}$ can be amalgamated in $\mathfrak{K}$ iff for every maximal congruence $\Theta\in C(\underline{B})$ there is a maximal congruence $\Phi\in C(\underline{C})$ with the same restriction $\Psi\in C(\underline{A})$ to $\underline{A}$ as Θ such that the amalgam $f/\Psi:\underline{A}/\Psi\rightarrowtail\underline{B}/\Theta$, $g/\Psi:\underline{A}/\Psi\rightarrowtail\underline{C}/\Phi$ can be amalgamated in the class $\mathfrak{K}_s$ of all simple $\mathfrak{K}$-algebras (and the same condition for $\underline{B}$ and $\underline{C}$ interchanged).

Proof: necessity: Let $h:\underline{B}\rightarrowtail\underline{D}$, $l:\underline{C}\rightarrowtail\underline{D}$ be an amalgamation of (f,g). For every maximal congruence $\Theta\in C(\underline{B})$ there is a maximal congruence $\Omega\in C(\underline{D})$, whose restriction to $\underline{B}$ is Θ, because of the CEP . Choose Φ and Ψ to be the restricton's of Ω to $\underline{C}$ and $\underline{A}$ respectively. Then we obviously have the amalgamation

$$\begin{array}{ccc} \underline{A}/\Psi & \longrightarrow & \underline{B}/\Theta \\ \downarrow & & \downarrow \\ \underline{C}/\Phi & \longrightarrow & \underline{D}/\Omega \end{array}$$

and all those algebras are simple.

sufficiency: Let

$$\begin{array}{ccc} \underline{A}/\Psi_i & \rightarrowtail & \underline{B}/\Theta_i \\ \downarrow & & \downarrow h_i \\ \underline{C}/\Phi_i & \xrightarrow[l_i]{} & \underline{D}_i \end{array}$$

be amalgamations of the $(f/\Psi_i, g/\Psi_i)$, where Θ_i runs over all maximal congruences of $\underline{B}$ and Φ_i runs over all maximal congruences of $\underline{C}$.(Here we also admit the universal congruence ∇ as maximal congruence).

Then because $\underline{B}$ and $\underline{C}$ are semisimple the natural mappings $h=(h_i|i\in I):\underline{B} \longrightarrow \prod_{i\in I} \underline{D}_i := \underline{D}$ and $l=(l_i|i\in I):\underline{C} \longrightarrow \underline{D}$ are embeddings and obviously amalgamate $f:\underline{A} \rightarrowtail \underline{B}$ and $g:\underline{A} \rightarrowtail \underline{C}$.

(2) COROLLARY : A discriminator variety $\mathcal{V}$ has the amalgamation property iff the class $\mathcal{O}_0$ of its simple non-singleton algebras has the amalgamation property.

(3) COROLLARY : A residually small discriminator variety has the amalgamation property iff it is generated by finitely many independent demi-semi-primal algebras.

(4) COROLLARY (QUACKENBUSH [127]): A variety generated by a quasi-primal algebra $\underline{A}$ has the amalgamation property iff $\underline{A}$ is demi-semi-primal.

Proof: It is clear that in the theorem (1) we can trivially choose $\Phi=\nabla$ if the restriction of Θ to $\underline{A}$ is ∇ , and then we have a trivial amalgamation with $\underline{D}=\underline{B}/\Theta$. So we only have to check the AP in the class of non-singleton simple algebras. This proves (2). To see (3) first observe that for an independent set of demi-semi-primal algebras the set of non-singleton subalgebras has the amalgamation property by 1.19 . Conversely 1.19 also says that for a finite set of quasi-primal algebras the set of non-singleton subalgebras has the AP iff the maximal algebras form an independent set of demi-semi-primal algebras. (4) now follows easily from (3). □

In the rest of this section we are going to give the discriminator varieties corresponding to the examples in §1 .

If we consider the varieties generated by the examples in 1.20 we of course get discriminator varieties, and the striking fact about all these varieties is that all members of these varieties have *central* elements (these are elements which define a direct decomposition of the algebra, just like central idempotents in rings or complemented elements in distributive lattices). The examples in 1.20 are then precisely those algebras which have only trivial central elements.

2.6 HEYTING ALGEBRAS

Most of the examples in 1.20 have an underlying Heyting-algebra or even more special an underlying Boolean algebra. So we start with a study of Heyting-algebras and their central elements in the above sense.

A Heyting-algebra $\underline{H} = (H;\wedge,\vee,\rightarrow,0,1)$ is a bounded distributive lattice with a relative pseudocomplementation

$x \leq a\rightarrow b$ iff $x\wedge a \leq b$.

It is well known that Heyting algebras can be equationally defined by the equations for distributive lattices together with the following three equations

(H1) $x\wedge y \rightarrow y = 1$

(H2) $x\wedge(x \rightarrow y) = x\wedge y$

(H3) $x\wedge(x\wedge y \rightarrow z) = x\wedge(y \rightarrow z)$

We are going to see that also for Heyting-algebras the central elements are the complemented elements (just as for distributive lattices). Let $a\in H$ be a complemented element and let $a'\in H$ be its complement.

$\underline{H}_a := ([0,a];\wedge,\vee,*,0,a)$, where $x*y := a\wedge(x\rightarrow y)$

$m_a: H \longrightarrow [0,a]$ is defined by $m_a(x) := a\wedge x$.

(1) m_a is a Heyting-homomorphism .

Proof: First note that in every Heyting-algebra the following equations hold:

(H4) $x\vee y \rightarrow z = (x \rightarrow z)\wedge(y\rightarrow z)$

(H5) $x \rightarrow y\wedge z = (x \rightarrow y)\wedge(x\rightarrow z)$

The inequality $a\wedge x\wedge a' = 0 \leq y$ implies $a \leq x\wedge a' \rightarrow y$.

$$\begin{aligned} m_a(x\rightarrow y) &= (x\rightarrow y)\wedge a = ((x\wedge a)\vee(x\wedge a')\rightarrow y)\wedge a = (x\wedge a \rightarrow y)\wedge(x\wedge a' \rightarrow y)\wedge a \\ &= (x\wedge a \rightarrow y)\wedge a = (x\wedge a \rightarrow y)\wedge(x\wedge a \rightarrow a)\wedge a = (x\wedge a \rightarrow y\wedge a)\wedge a \\ &= m_a(x)*m_a(y) \ . \ \square \end{aligned}$$

As we have just proved m_a and $m_{a'}$ are Heyting-homomorphism, and as we know from distributive lattices the define a direct decomposition of $\underline{H}$ into the Heyting-algebras $\underline{H}_a$, $\underline{H}_{a'}$. This direct decomposition of $\underline{H}$ we use in the sequel in the examples.

For the first case we consider the Heyting-algebras generated by the examples 1.20(1) and 1.20(6), where we have an additional operation d or $^+$ respectively. In example 1.20(1) we define $x^+ := d(x) \rightarrow 0$, then $d(x) = x^{++}$.

$^+$-HEYTING-ALGEBRAS

We consider the equational class $\mathcal{H}_+$ of all algebras $\underline{H} = (H;\wedge,\vee,\rightarrow,^+,0,1)$ which satisfy

(H0) $(H;\wedge,\vee,0,1)$ is a bounded distributive lattice

(H1) $x\wedge y \rightarrow y = 1$

(H2) $x\wedge(x \rightarrow y) = x\wedge y$

(H3) $x\wedge(x\wedge y \rightarrow z) = x\wedge(y \rightarrow z)$

(H1–H3): $\rightarrow$ is a relative pseudocomplementation

(H1$^+$) $1^+ = 0$

(H2$^+$) $x\vee x^+ = 1$

(H3$^+$) $x\vee(x\vee y)^+ = x\vee y^+$

(H1$^+$–H3$^+$): $^+$ is a dual pseudocomplementation

(S$^+$) $x^+\wedge x^{++} = 0$ dual Stone-identity

(2): The equations (H0,H1$^+$,H2$^+$,H3$^+$,S$^+$) imply the equation

(H) $(x\wedge y^+)^+\wedge y^+ = x^+\wedge y^+$

Proof: We first show the equation $(x\wedge y)^+ = x^+\vee y^+$. Observe that $x^+ \leq a$ iff $x\vee a = 1$ and that $^+$ is antitone, so trivially $(x\wedge y)^+ \geq x^+\vee y^+$ holds. Now $(x\wedge y)^+ \leq x^+\vee y^+$ iff $1 = (x\wedge y)\vee x^+\vee y^+$ $= ((x\vee x^+)\wedge(y\vee x^+))\vee y^+ = y\vee x^+\vee y^+ = 1$.
We use this as follows: $(x\wedge y^+)^+\wedge y^+ = (x^+\vee y^{++})\wedge y^+ =$ $= (x^+\wedge y^+)\vee(y^{++}\wedge y^+) = x^+\wedge y^+$ by (S$^+$). □

The value of the equation (H) is that it says that the mapping $m_{y^+} : H \rightarrow [0,y^+]$. is also a $^+$-homomorphism if one defines on $[0,y^+]$ $x^\bullet := x^+\wedge y^+$.

(3) The variety $\mathcal{H}_+$ is a discriminator variety and the directly indecomposable algebras in $\mathcal{H}_+$ are precisely those from 1.20(6).

(4) The variety $\mathcal{H}_d$ which is defined as $\mathcal{H}_+$ only (H1$^+$,H2$^+$, H3$^+$) are replaced by their consequences (H) and

(H6) $x^+\vee x^{++} = 1$

(H7) $x \geq x^{++}$

is a discriminator variety and the directly indecomposable members of $\mathcal{H}_d$ are precisely those from 1.20(1).

Proof: For each $x\in H$ x^+ is complemented by (H6) and defines a direct decomposition by (H), so if $\underline{H}$ is directly indecomposable, x^+ can only take the values 0 or 1. (H7) forces $^+$ to act as follows: $x^+ = \begin{cases} 0 & \text{if } x=1 \\ 1 & \text{if } x\neq 1 \end{cases}$. $^+$ is the dual pseudocomplementation on $\underline{H}$ iff 1 is $\vee$-irreducible. □

A very important subvariety of the variety $\mathcal{H}_+$ of all dually Stone-Heyting-algebras is the variety $\mathcal{RS}_+$ of all dually Stone relatively Stone algebras, which is defined by the additional identity

(RS) $(x\to y)\vee(y\to x) = 1$ (relative Stone identity)

(5) $\mathcal{RS}_+$ is a discriminator variety and the directly indecomposable algebras in $\mathcal{RS}_+$ are precisely the chains in example 1.20(6) .

Proof: As $x\le y \iff x\to y=1$, the equation $(x\to y)\vee(y\to x)=1$ says that the underlying lattice is a chain, whenever 1 is $\vee$-irreducible. □

2.7 BOOLEAN ALGEBRAS

Boolean algebras are easily seen to be a discriminator variety because if a Boolean algebra contains an element $x\neq 0,1$, then $\underline{B} \cong [0,x]\times[0,x']$ is a non-trivial direct decomposition , so the variety $\mathcal{B}$ of Boolean algebras is generated by the algebra $\underline{2}$ in example 1.20(2), hence $\mathcal{B}$ is a discriminator variety.

The variety $\mathcal{R}$ of relatively complemented distributive lattices (=generalized Boolean algebras) is generated by the 2-element algebra $\underline{R}_2$ in 1.20(3) because if $\underline{R}\in\mathcal{R}$ has an element $x\neq 0,1$, then $\underline{R} \cong (x]\times[x)$ and thus $\underline{R}_2$ is the only directly indecomposable member of $\mathcal{R}$ and $\mathcal{R}$ is a discriminator variety.

2.8 ŁUKASIEWICZ-ALGEBRAS

The variety $\mathcal{L}_n$ of Łukasiewicz-algebras of order n is the class of algebras $\underline{L} = (L;\wedge,\vee,^-,D_1,\dots,D_{n-1},0,1)$ satisfying

(L0) $(L;\wedge,\vee,0,1)$ is a bounded distributive lattice

(L1) $\bar{\bar{x}} = x$, $\overline{x\wedge y} = \bar{x}\vee\bar{y}$ (De Morgan laws)

(L2) $D_i(x\wedge y) = D_i(x)\wedge D_i(y)$ $1\le i<n$

$D_i(x\vee y) = D_i(x)\vee D_i(y)$ $1\le i<n$

(L3) $D_i(x) \ge D_j(x)$ $1\le i\le j<n$

(L4) $D_i(x)\vee\overline{D_i(x)} = 1$ $1\le i<n$

$D_i(x)\wedge\overline{D_i(x)} = 0$ $1\le i<n$

(L5) $D_i(\bar{x}) = \overline{D_{n-i}(x)}$ $1\le i<n$

(L6) $D_i(D_j(x)) = D_j(x)$ $\qquad 1 \le i,j < n$

(L7) $D_{n-1}(x) \le x \le D_1(x)$

(L8) $\bar{x} \wedge D_{n-1}(x) = 0$

$\bar{x} \vee D_1(x) = 1$

(L9) $D_i(0) = 0$, $D_i(1) = 1$ $\qquad 1 \le i < n$

(L10) $y \le x \vee \overline{D_i(x)} \vee D_{i+1}(y)$ $\qquad 1 \le i \le n-2$

From BALBES-DWINGER [10] one gets that the operation

$$x \to y := y \vee \bigwedge\{\overline{D_i(x)} \vee D_i(y) \mid 1 \le i < n\}$$

is the relative pseudocomplementation on L, and a one-line proof shows that $\to$ satisfies the relative Stone identity (RS)

$$(x\to y) \vee (y \to x) = x \vee y \vee \bigwedge\{\overline{D_i(x)} \vee D_i(y) \vee \overline{D_j(y)} \vee D_j(x) \mid 1 \le i,j < n\} = 1.$$

An evenly short proof shows that the operation

$$x^+ := D_1(\bar{x})$$

is a dual pseudocomplementation and satisfies the dual Stone identity (S^+) :

$(H1^+)$: $1^+ = D_1(\bar{1}) = D_1(0) = 0$

$(H2^+)$: $x \vee x^+ = x \vee D_1(\bar{x}) = 1$

$(H3^+)$: $x \vee (x \vee y)^+ = x \vee D_1(\overline{x \vee y}) = x \vee (D_1(\bar{x}) \wedge D_1(\bar{y})) = x \vee D_1(\bar{y}) = x \vee y^+$

(S^+) : $x^+ \wedge x^{++} = D_1(\bar{x}) \wedge D_1(\overline{D_1(\bar{x})}) = D_1(\bar{x}) \wedge \overline{D_1(\bar{x})} = 0$

Altogether we have shown:

(1) $(L; \wedge, \vee, \to, {}^+, 0, 1)$ is in the variety $\mathcal{RS}_+$.

(2) If we define on $[0,a^+]$ new operations $', D_1', \ldots, D_{n-1}'$ by $x' := \bar{x} \wedge a^+$ and $D_i'(x) := D_i(x) \wedge a^+$, then the mapping $m_{a^+} : L \longrightarrow [0,a^+]$ becomes a homomorphism.

Proof: $m_{a^+}(\bar{x}) = \bar{x} \wedge a^+ = (\bar{x} \vee \overline{a^+}) \wedge a^+ = \overline{(x \wedge a^+)} \wedge a^+ = (m_{a^+}(x))'$

$m_{a^+}(D_i(x)) = D_i(x) \wedge a^+ = D_i(x) \wedge D_i(a^+) \wedge a^+ = D_i'(m_{a^+}(x)$.

(3) The variety of Łukasiewicz-algebras of order n is a discriminator variety and the directly indecomposable members of $\mathcal{L}_n$ are precisely the subalgebras of the example 1.20(4) .

Proof: From 2.6 we know already that the indecomposables are chains and we have to show that they are embeddable into $\underline{L}_n$. If $\underline{C}$ is this chain we define $f: \underline{C} \longrightarrow \underline{L}_n$ by $f(x) := i$ iff $D_j(x) = 0 \iff j > i$. f is easily seen to be a 0-1-homomorphism, so as $\underline{C}$ is simple it is an embedding.

Post-algebras of order n can be considered as Łukasiewicz-algebras with n distinguished constants and so we get

(4) The variety $\mathcal{P}_n$ of Post-algebras of order n is a discriminator variety and the only subdirectly irreducible member of $\mathcal{P}_n$ is the algebra $\underline{P}_n$ in 1.20(5) .

2.9 CYLINDRIC ALGEBRAS

The variety $\mathcal{C}_n$ of cylindric algebras of dimension n is the class of all algebras $\underline{C} = (C;\wedge,\vee,',0,1,c_1,\ldots,c_n)$ with

(C0) $(C;\wedge,\vee,',0,1)$ is a Boolean algebra

(C1) $c_k0 = 0$

(C2) $x \leq c_kx$

(C3) $c_k(x\wedge c_ky) = c_kx \wedge c_ky$

(C4) $c_kc_jx = c_jc_kx$ $\qquad 1\leq k,j \leq n$

The cylindric algebras of dimension 1 are called monadic algebras and it is known (MONK [107]) that the lattice of subvarieties of $\mathcal{C}_1$ is a chain isomorphic to $\omega+1$.

(1) Every cylindric algebra satisfies the following conditions

(C5) $c_k1 = 1$

(C6) $c_kc_kx = c_kx$

(C7) $x\leq y \Rightarrow c_kx\leq c_ky$

(C8) $c_ka = a \Rightarrow c_k(x\wedge a) = c_kx \wedge a$

(C9) $c_ka = a \Rightarrow c_ka' = a'$ $\qquad 1\leq k\leq n$

Proof: (C5) follows from (C2) and (C6) from (C3) and (C5).
(C7) : $x\leq y \Rightarrow x\leq c_ky \Rightarrow c_kx\wedge c_ky=c_k(x\wedge c_ky)=c_kx$
(C8) : $c_k(x\wedge a) = c_k(x\wedge c_ka) = c_kx\wedge c_ka = c_kx\wedge a$
(C9) : $c_ka'\wedge a = c_k(a'\wedge a) = c_k0 = 0$, hence $c_ka'\leq a'$. By (C2) we now get $c_ka' = a'$. □

(2) If $\underline{C}\in\mathcal{C}_n$ and $a\in\underline{C}$ such that $c_ka=a$ for all $k=1,\ldots,n$, then $[0,a]$ with the Operations $\wedge,\vee,\_'\wedge a,0,a,c_1,\ldots,c_n$ becomes an algebra in $\mathcal{C}_n$ and $m_a:\underline{C}\longrightarrow[0,a]$ becomes a homomorphism.

Proof: If $x\leq a$ then $c_kx \leq c_ka = a$ by (C7), so $c_1,\ldots,c_n$ are operations on $[0,a]$. (C8) says that m_a is a homomorphism. □

(3) COROLLARY : The variety $\mathcal{C}_n$ of cylindric algebras of dimension n is a discriminator variety and the directly indecomposable members of $\mathcal{C}_n$ are precisely the algebras in example 1.20(7).

Proof: From (C4) and the fact that each c_k is a closure operator (C2,C7,C6) it immediately follows that

$(\forall k\in\{1,..n\}\ c_k a = a) \Leftrightarrow (a = c_1 c_2 \ldots c_n(a)) \Leftrightarrow (\exists x\ a = c_1 c_2 \ldots c_n(x))$.

As each a with this property defines a direct decomposition by (2) and (C9) an algebra $\underline{C} \in \mathcal{C}_n$ is directly indecomposable iff

$c_1 c_2 \ldots c_n(x) = \begin{cases} 1 & \text{if } x \neq 0 \\ 0 & \text{if } x = 0 \end{cases}$. □

(4) COROLLARY : The only proper subvarieties of the variety $\mathcal{C}_1$ of monadic algebras are the varieties $\mathcal{M}_n$ generated by the simple monadic algebra $\underline{M}_n = (2^n;\wedge,\vee,',0,1,c_1)$,where $c_1(x) = \begin{cases} 1 & \text{if } x \neq 0 \\ 0 & \text{if } x = 0 \end{cases}$.

Proof: The subdirectly irreducible monadic algebras are precisely the Boolean algebras with the above operation c_1 . Obviously each Boolean subalgebra of such a monadic algebra is also such a monadic algebra, so the only residually small subvarieties of $\mathcal{C}_1$ are precisely the varieties $\mathcal{M}_n$ above. If $\underline{B}$ is any infinite Boolean algebra, then every other Boolean algebra can be embedded into some ultrapower of $\underline{B}$, so each infinite simple monadic algebra generates the whole variety $\mathcal{C}_1$. □

2.10 RELATION ALGEBRAS

Relation algebras were introduced by A.TARSKI [162] in 1941 in order to investigate the calculus of (binary) relations with respect to the operations $\wedge$ (intersection), $\vee$ (union), ' (complementation), $\circ$ (relational product), $^+$ (inversion) 0 (empty relation), 1 (universal relation), and Δ (identity relation). Relation algebras are defined as algebras $\underline{R} = (R;\wedge,\vee,',0,1,\circ,^+,\Delta)$ satisfying the following axioms:

(R0) $(R;\wedge,\vee,',0,1)$ is a Boolean algebra

(R1) $x\circ(y\circ z) = (x\circ y)\circ z$, $x\circ\Delta = x$

(R2) $x^{++} = x$, $(x\circ y)^+ = y^+ \circ x^+$, $(x\vee y)^+ = x^+ \vee y^+$

(R3) $(x\vee y)\circ z = (x\vee z)\circ(y\vee z)$

(R4) $x\circ(x^+ \circ y)' \leq y'$

In relation algebras the so called ideal elements play the rôle that the complemented elements do for Heyting-algebras. they were investigated in CHIN-TARSKI [26] . $a\in R$ is called *ideal* iff $1\circ a\circ 1 = a$. It is easy to see from the axioms that the mapping $x \mapsto 1\circ x\circ 1$ is a closure operator on R, which assigns to each element of R the smallest ideal element containing it. For an ideal element $a\in R$ in [26] the following equations were shown:

(1) $a' = 1\circ a'\circ 1$ (the complement of an ideal element is ideal)

(2) $(x\wedge a)^{+} = x^{+}\wedge a$ (m_a is a $^{+}$-homomorphism)

(3) $(x\circ y)\wedge a = (x\wedge a)\circ(y\wedge a)$ (m_a is a $\circ$-homomorphism)

(4) COROLLARY : The variety $\mathfrak{Re}$ of relation algebras is a discriminator variety and the directly indecomposable members of $\mathfrak{Re}$ are precisely the examples 1.20(9) .

2.11 ARITHMETICAL RINGS

In a ring we have a situation similar to that in the Heyting-algebras before, namely the central idempotent elements in a ring play the same rôle as the complemented elements in a Heyting-algebra. We are interested in discriminator varieties of rings, so each ring in question must have a distributive ideal lattice, in other words, we are interested in varieties of arithmetical rings. In [106] MICHLER-WILLE have shown that each variety of arithmetical rings is generated by finitely many finite fields. This proof uses heavy machinery from ringtheory, so we give a more elementary proof, which in this form is due to A.F.Pixley.

THEOREM (MICHLER-WILLE) : For a variety $\mathfrak{R}$ of rings t.f.a.e.

(1) $\mathfrak{R}$ is generated by finitely many finite fields.

(2) $\mathfrak{R}$ is a discriminator variety of rings.

(3) $\mathfrak{R}$ is a variety of arithmetical rings.

(4) There is a polynomial $t(x)$ with integer coefficients and without constant term such that each $\underline{R}\in\mathfrak{R}$ satisfies the equation $x\cdot t(x) = x$.

(5) There is an $n\geq 2$ such that each $\underline{R}\in\mathfrak{R}$ satisfies $x^n = x$.

Proof: (1)=>(2)=>(3) are clear from 1.20(11) and 2.2 so we prove (3)=>(4)=>(5)=>(1) and we use the obvious implications (1)=>(5)=>(4) on the way. (3)=>(4) is the same as in [106] . By Pixley's characterization of arithmetical varieties 1.11 we have a ternary polynomial $s(x,y,z)$ with integer coefficients satisfying the equations $s(x,x,y) = x = s(x,y,x) = s(y,x,x)$. These equations imply that the polynomial $p(x,y,z) :=$ $s(x,y,z)-s(0,0,z)-s(0,y,0)-s(x,0,0)+2s(0,0,0)$ is a polynomial with integer coefficients without constant or linear term, satisfying the same equations as s does. Thus there is a polynomial $t(x)$ as desired in (4) satisfying $x \cdot t(x) = p(x,x,x)$.

(4)$\Rightarrow$(1,5): (i) $x^2 = 0 \Rightarrow x \cdot t(x) = 0 \Rightarrow x = 0$
(ii) $e^2 = e \Rightarrow ex = xe$, because $(ex - exe)^2 =$ $= exex - exexe - exeex + exeexe = 0$, hence $ex = exe$ and similarly $xe = exe$. (iii) $t(x) \cdot t(x) = t(x)$ is an easy consequence of $x \cdot t(x) = x$. Obviously we have (iv) $t(x) = 0 \Rightarrow x = 0$.
We have shown so far that for each $x \neq 0$ $t(x)$ is a central idempotent $\neq 0$ in $\underline{R}$ so the directly indecomposable rings in $\mathcal{R}$ are characterized by $t(x) = \begin{cases} 1 & \text{if } x \neq 0 \\ 0 & \text{if } x = 0 \end{cases}$, in particular they are rings with unit. Thus there is a polynomial $q(x)$ possibly with a constant term such that $t(x) = x \cdot q(x)$, and if $x \neq 0$ we have $x \cdot q(x) = 1$, so $\underline{R}$ is a skew field, and moreover satisfies the equation $x \cdot t(x) - x = 0$. Thus $\underline{R}$ has at most $\deg(x \cdot t(x))$ many elements, and by Wedderburn's theorem $\underline{R}$ is a finite field with at most $\deg(x \cdot t(x))$ elements. □

2.12 RINGS WITH AN OPERATOR

In the examples 1.20(10,12,13,14) we considered rings with an additional operator (=unary operation). The most general case is the class of rings $\underline{R}$ with an operator $g: R \to R$ which assigns to each element $x \in R$ a central idempotent $g(x)$ satisfying $g(x) \cdot x = x$.

(1) THEOREM : The variety $\mathcal{R}_g$ of algebras $\underline{R} = (R;+,-,0,\cdot,g)$ satisfying

(R0) $(R;+,-,0,\cdot)$ is an associative ring

(R1) $g(x) \cdot y = y \cdot g(x)$, $g(x) \cdot g(x) = g(x)$

(R2) $g(x) \cdot x = x$

(R3) $g(x \cdot g(y)) = g(x) \cdot g(y)$
$g(x - x \cdot g(y)) = g(x) - g(x) \cdot g(y)$

is a discriminator variety and the directly indecomposable algebras in $\mathcal{R}_g$ are precisely the examples 1.20(10) .

As special cases we get the biregular rings, the Baer *-rings, and the strongly regular rings.
A ring is called *biregular* iff for each $x \in R$ there is a central idempotent $x^+ \in R$ such that x^+ generates the same principal ideal of R as x does, Obviously x^+ is uniquely determined by x, so we can consider $^+$ a unary operation.
A ring $\underline{R}$ is called a *Baer *-ring* iff for each $x \in R$ the annihilator ideal of x is generated by a central idempoetent x^* (which is uniquely determined by x).
A ring $\underline{R}$ is called *strongly regular* iff for each element $x \in R$ there exists an element $x^{-1} \in R$ such that $x^2 \cdot x^{-1} = x$. x^{-1} is uniquely determined by x except for $x=0$ where we define $0^{-1} := 0$. Note that each ring-homomorphism preserves the so defined operators $^+$,*, and $^{-1}$, so the homomorphic images of biregular rings, Baer *-rings, strongly regular rings are again biregular, Baer *,or srongly regular respectively.
The directly indecomposable members are therefore precisely the examples 1.20(12,13,14).

In the sequel we give equational bases for the Baer *-rings and for the strongly regular rings. Unfortunately the biregular rings do not form a variety even by adding new operations, because the property of being simple is not first order expressible for a ring. In the commutative case however biregular and strongly regular rings coincide, because the subdirectly irreducibles are precisely the fields.

(2) THEOREM : The class $\mathcal{RB}^*$ of Baer *-rings can be equationally defined by

(RB0) $(R;+,-,0,\cdot)$ is an associative ring

(RB1) $x^* \cdot y = y \cdot x^*$, $x^* \cdot x^* = x^*$

(RB2) $x \cdot x^* = 0$, $x \cdot 0^* = x$

(RB4) $(x(xy)^*)^* \cdot y(yx)^* = y(yx)^*$

In fact (RB2,3) are equivalent to

$(*)$ $xy = 0 = yx \iff x^*y = y$.

$\mathcal{RB}^*$ is a discriminator variety and the directly indecomposable members of $\mathcal{RB}^*$ are precisely the examples 1.20(13).

Proof: It is clear that (RB0,1) and $(*)$ axiomatize Baer *-rings, so we have to show $(*) \iff$ (RB2,3).

$(*) \Rightarrow$ (RB2,3) $x^* \cdot x^* = x^* \Rightarrow x \cdot x^* = 0$

$x0 = 0 = 0x \Rightarrow 0^*x = x$

$$x(xy)^* \cdot y(yx)^* = xy(xy)^*(yx)^* = 0$$
$$y(yx)^* \cdot x(xy)^* = yx(yx)^*(xy)^* = 0 \quad \text{implies:}$$
$$(x(xy)^*)^* \cdot y(yx)^* = y(yx)^*$$

(RB2,3)=>(*) $x^*y = y \Rightarrow xy = xx^*y = 0 = yxx^* = x^*yx = yx$

$$xy = 0 = yx \Rightarrow x^*y = (x0^*)^* \cdot y0^* = (x(xy)^*)^* y(yx)^* = y(yx)^* = y0^* = y. \quad \square$$

(3) THEOREM : The class $\mathfrak{RS}$ of strongly regular rings can be equationally defined by the axioms:

(RS0) $(R;+,-,0,\cdot)$ is an associative ring

(RS1) $x^2 \cdot x^{-1} = x$, $0^{-1} = 0$

(RS2) $(xy)^{-1} = y^{-1} \cdot x^{-1}$

(RS3) $(y - yxx^{-1})^{-1} = y^{-1} - y^{-1}xx^{-1}$.

$\mathfrak{RS}$ is a discriminator variety and the directly indecomposables of $\mathfrak{RS}$ are precisely the examples 1.20(14).

Proof: We show that xx^{-1} is a central idempotent, then the theorem follows because the equations (RS2,3) say that the direct decomposition defined by xx^{-1} is also a decomposition with respect to the operation $^{-1}$. By (RS1) we have $x^2=0 \Rightarrow x=0$, so as in 2.11 we have that each idempotent is central. $(xx^{-1}x - x)^2 = xx^{-1}xxx^{-1}x - xx^{-1}xx - xxx^{-1}x + xx = 0$, hence $xx^{-1}x = x$ and thus xx^{-1} is idempotent. $\square$

2.13 COMPLEMENTARY SEMIGROUPS

Complementary semigroups (or divisibility semigroups) were introduced and investigated in BOSBACH [16] , and later on [17 - 20] generalized to the non-commutative case, but we restrict ourselves to the commutative case here.

The author is very indebted to Prof. Bruno Bosbach for providing him with these examples of discriminator varieties and with the proofs needed .

A commutative semigroup $(S,\cdot)$ is called *complementary* if for all $a,b \in S$ there is a unique element $a*b \in S$ such that $ax \in bS \Leftrightarrow x \in (a*b)S$. If we write $x \leq y$ for "x divides y" , the above condition reads $b \leq ax \Leftrightarrow a*b \leq x$.
By definition $*$ is a binary operation on a complementary semigroup and from BOSBACH [16] we can collect the following facts about complementary semigroups.

(1) $(S;\cdot,*)$ is a complementary semigroup iff the following equations hold

(CS0) $xy = yx$

(CS1) $x(y*y) = x$

(CS2) $x(x*y) = y(y*x)$

(CS3) $x*(y*z) = xy*z$

(2) $(S;\leq)$ is a join-semilattice with the join $x \cup y := x(x*y)$ and the smallest element $1 = x*x$, which is the unit of the semigroup $(S;\cdot)$.

(3) If $(S;\leq)$ is also a lattice (the meet denoted by $\cap$), then the following conditions hold

(i) $x(y \cap z) = xy \cap xz$

(ii) $x*(y \cap z) = x*y \cap x*z$

(iii) $(x \cap y)*z = x*z \cup y*z$

(iv) $e \cdot e = e \Rightarrow e \cup x = ex$

(v) $a \cap b = 1 \;\&\; a \leq bc \Rightarrow a \leq c$

We are particularly interested in the *normal* complementary semigroups which are defined by the additional equation

(CSN) $(x*y)*y = (y*x)*x$.

In the normal complementary semigroups $(S;\leq)$ is a lattice and the meet is given by $x \cap y := (x*y)*y$ (see BOSBACH [18]). It is easy to see that the examples in 1.20(15) are normal. In the normal complementary semigroups the idempotent elements $e = e^2$ play an important role. For every idempotent $e = e^2$ we have

(4) (i) $e \cap (e*x) = 1$

(ii) $(e \cap x)(e*x) = x$

(iii) $e \cap x = 1 \Leftrightarrow e*x = x$

(iv) $e \cap x = x \Leftrightarrow e*x = 1$

(v) $e \cap xy = (e \cap x)(e \cap y)$

(vi) $e*xy = (e*x)(e*y)$

(vii) $e \cap (x*y) = (e \cap x)*(e \cap y)$

(viii) $e*(x*y) = (e*x)*(e*y)$

<u>Proof:</u> (i): $e \cap (e*x) = (e*(e*x))*(e*x)$ = by (CS3)

$= (e^2*x)*(e*x) = (e*x)*(e*x) = 1$

(ii): $(e \cap x)(e*x) = e(e*x) \cap x(e*x)$ = by (e,iv)

$= ex \cap x(e*x)$ = by (3,i)

$= x(e \cap (e*x)) = x$ by (4,i)

(iii) $e \cap x = 1 \Rightarrow e*x = x$ by (4,ii) ; $e*x = x$ implies $e \cap x = (e*x)*x = x*x = 1$.

(iv) is is obvious as $e \cap x = x \Leftrightarrow x \leq e \Leftrightarrow e*x = 1$.

(v) $e \cap xy = e^2 \cap ex \cap ey \cap xy$ (as $e^2 = e \leq ex, ey$) = by (CS0) and (3,i)
$= (e \cap x)(e \cap y)$

(vi): $(e \cap xy)(e * xy) = xy = (e \cap x)(e \cap y)(e * x)(e * y)$ by (4,ii)
$(e \cap xy) = (e \cap x)(e \cap y)$ by (4,v)
$e \cap xy \cap (e * xy) = 1$ by (4,i)
$(e \cap x)(e \cap y) \cap (e * x)(e * y) \leq e^2 \cap (e * x)(e * y) = e^2 \cap e(e * x) \cap e(e * y) \cap$
$\cap (e * x)(e * y) = [e \cap (e * x)][e \cap (e * y)] = 1$

Now use (3,v) to cancel $e \cap xy$ in the above equation and to get $e * xy = (e * x)(e * y)$.

(vii): $e \cap (x * y) = (e * (x * y)) * (x * y) =$ by (CS3)
$= (xe * y) * (x * y) =$ by (3,iv)
$= (x(x * e) * y) * (x * y) =$ by (CS3),(CS0)
$= ((x * e) * (x * y)) * (x * y) = (x * e) \cap (x * y) =$ by (3,ii)
$= x * (e \cap y) = 1 \cup (x * (e \cap y)) =$ by (4,iv)
$= (e * (e \cap y)) \cup (x * (e \cap y)) =$ by (3,iii)
$= (e \cap x) * (e \cap y)$

(viii): $e * (x * y) = ex * y = e(e * x) * y = (e * x) * (e * y)$
by (CS3),(3,iv),(CS0), and again (CS3) . □

Combining all these results we have

(5) If e is an idempotent element in the normal complementary semigroup $(S; \cdot, *)$, then the mappings $\phi, \psi : S \longrightarrow S$, $\phi x := e \cap x$ and $\psi x := e * x$ are nomomorphisms and they define a direct decomposition of S into its subalgebras $\phi S = \{x \mid e \cap x = x\}$ and $\psi S = \{x \mid e \cap x = 1\} = \{x \mid e * x = x\}$.

The examples in 1.20(15) have in common the property that they are of finite rank , which means that there is a $k \geq 1$ such that each element of the semigroup $(S; \cdot)$ is nilpotent of rank at most k . In particular these examples satisfy the equation

(Rk) $x^k = x^{k+1}$.

We define the variety $\mathcal{CS}_k$ of all *complementary semigroups of rank k* by the equations (CS0),(CS1),(CS2),(CS3),(CSN), and (Rk).

(6) THEOREM : The variety $\mathcal{CS}_k$ of all complementary semigroups of rank k is a discriminator variety and the subdirectly irreducible members of $\mathcal{CS}_k$ are precisely the algebras $\underline{S}_n$ $(n \leq k+1)$ from example 1.20(15) .

Proof : The equation (Rk) says that for each $x \neq 1$ x^k is an idempotent different from 1 , so the directly indecomposable members of $\mathcal{CS}_k$ have a zero-element 0 and $x^k = \begin{cases} 0 & \text{if } x \neq 1 \\ 1 & \text{if } x = 1 \end{cases}$,

so the same polynomial t as in 1.20(15) serves as the discriminator. It remains for us to show that if $\underline{S}$ is a directly indecomposable member of $\mathfrak{S}_k$, $\underline{S}$ is isomorphic to some $\underline{S}_n$. From BOSBACH [18] we know that all subdirectly irreducible normal complementary semigroups are chains with respect to $\leq$. Let $n = \min\{i \leq k \mid x \in S \Rightarrow x^i = x^{i+1}\}$ and let $a \in S$ be such that $a^{n-1} \neq a^n = a^{n+1}$. It suffices to show $S = \{1, a, a^2, \ldots, a^n\}$. Assume $s \in S \setminus \{a^0, a^1, \ldots, a^n\}$, where $a^0 := 1$.
As S is a chain there is an $i < n$ such that $a^i < s < a^{i+1}$. Let $b := a^i * s$. If $a \leq b$ we get $s = a^i \cup s = a^i(a^i * s) \leq a^{i+1}$, a contradiction, thus $b < a$. Let c equal b or $b * a$, whichever is smaller. Then $c^2 \leq b(b*a) = b \cup a = a$ and thus $c^{2i-1} < c^{2i} \leq a^i$ $(i = 1, 2, \ldots, n)$, which contradicts the choice of n. □

§ 3 SHEAVES OVER BOOLEAN SPACES

This section just gives the technical details on sheaves of algebras over Boolean spaces which we are going to use later. The use of sheaf-representations in various parts of algebra have become popular in recent years and, just to name a few, we have sheaf-representations for rings (AHRENS-KAPLANSKY [3] DAVIS [36] , DAUNS-HOFMANN [33] , HOFMANN [63] , KIST [89] , MEWBORN [105] , MULVEY [108] , PEERCY [111] , and PIERCE [112]) semigroups (ADAMS [2] , KEIMEL [83,85]), ℓ-groups(KEIMEL [84]) f-rings (KEIMEL [85,86,87]), distributive lattices (DAVEY [34]), and for universal algebras (COMER [30] , DAVEY [35] , KEIMEL [83] , KEIMEL-WERNER [88] , and WOLF [180]).

These representations also have lead to many model theoretic investigations for algebras of sections over a sheaf (see BULMAN FLEMING-WERNER [21] , BURRIS-WERNER [24] , COMER [31] , LIPSHITZ-SARACINO [98] , MACINTYRE [101]).

In order to keep this chapter self-contained we repeat the definitions of a sheaf of algebras and its algebra of (global) sections and we also give the standard construction of a sheaf of algebras from a subdirect representation of an algebra. In the sequel we will only consider sheaves over Boolean spaces, so we immediately turn to this special case.

3.1 SHEAVES OF ALGEBRAS

A *sheaf of algebras over a topological space X* is a triple $\underset{\sim}{S} = (S,\eta,X)$, where

(S1) S and X are topological spaces (called the *sheaf-space* and the *base-space* of $\underset{\sim}{S}$)

(S2) $\eta:S \longrightarrow X$ is a local homeomorphism (i.e. for every point $s\in S$ there exists a neighborhood N of s in S such that ηN is open and $\eta: N \longrightarrow \eta N$ is a homeomorphism)

(S3) For every $x\in X$ the set $S_x := \{s\in S \mid \eta s = x\}$ forms an algebra which is called the *stalk* $\underline{S}_x$ over $x\in X$, and all stalks are of the same type.

(S4) For every n-ary $(n\geq 0)$ fundamental operation f on the stalks $\underline{S}_x$ $(x\in X)$ the induced mapping $f:S^{<n>} \longrightarrow S$ is continuous, where $S^{<n>} := \{(s_1,\ldots,s_n)\in S^n \mid \eta s_1 = \ldots = \eta s_n\}$ is endowed with the relative topology induced by the product topology on S^n

3.2 SECTIONS OF SHEAVES

Let $\underset{\sim}{S}=(S,\eta,X)$ be a sheaf of algebras.

A continuous map $\sigma:X \longrightarrow S$ is called a *section* of $\underset{\sim}{S}$ iff $\eta\circ\sigma = id_X$. $\Gamma\underset{\sim}{S}$ denotes the set of all sections of $\underset{\sim}{S}$; it forms a subalgebra of $\prod_{x\in X} \underline{S}_x$.

A very easy observation for sections of a sheaf is that *equalizers* are open, i.e.

(1) $\sigma,\tau\in\Gamma\underset{\sim}{S} \Rightarrow Eq(\sigma,\tau) := \{x\in X \mid \sigma x = \tau x\}$ is an open subset of X

For the proof let $\sigma x = \tau x$ and let N be a neighborhood of σx , on which η is a homeomorphism. Then σ and τ coincide on the open neighborhood $\sigma^{-1}N \cap \tau^{-1}N$ of x. □

3.3 THE STANDARD SHEAF

Let $\underline{R}$ be an algebra and $\{f_i:\underline{R} \twoheadrightarrow \underline{S}_i \mid i\in I\}$ a subdirect representation of $\underline{R}$, i.e. the f_i are all onto and the canonical homomorphism $f:\underline{R} \longrightarrow \prod_{i\in I} \underline{S}_i$ is one-to-one , or equivalently $\bigcap\{Ker f_i \mid i\in I\} = \Delta$.

The following construction leads to a representation of $\underline{R}$ by an algebra of sections over a sheaf of algebras.

(i) Endow I with a topology such that all sets $\{i \in I \mid f_i x = f_i y\}$, $x,y \in R$, become open subsets of I.

(ii) Define S to be the disjoint union $\cup\{S_i \mid i \in I\}$ of all algebras S_i and define $\eta: S \longrightarrow I$ to be the obvious mapping.

(iii) Endow S with a topology such that all sets $\{f_i x \mid i \in U\}$, where $x \in R$ and U open in I , become open subsets of S .

(iv) For every $x \in R$ define $\hat{x}: I \longrightarrow S$ by $\hat{x}(i) := f_i(x)$.

Then we get the following results

(1) $\underset{\sim}{S} = (S,\eta,X)$ is a sheaf of algebras over I with the algebras $\underline{S}_i$ as stalks over $i \in I$.

(2) For every $x \in R$ the mapping $\hat{x}: I \longrightarrow S$ is a section.

(3) The mapping $\hat{}: \underline{R} \longrightarrow \Gamma\underset{\sim}{S}$ is an injective homomorphism.

We call $\hat{}: \underline{R} \longrightarrow \Gamma\underset{\sim}{S}$ the *standard sheaf representation* of $\underline{R}$ with respect to the subdirect representation $\{f_i: \underline{R} \longrightarrow \underline{S}_i \mid i \in I\}$ and $\underset{\sim}{S}$ is called the *standard sheaf* of $\underline{R}$.
A very good exposition of the proofs can be found in DAVEY [35]

We now turn to the investigation of sheaves of algebras over a Boolean space. It turns out that the topology on the base space and on the sheaf space is uniquely determined by the sections, provided the sheaf space is Hausdorff and the sheaf has at most one 1-element stalk. In the sequel we call a sheaf $\underset{\sim}{S} = (S,\eta,X)$ *Boolean* , iff X is a Boolean space, S is a Hausdorff-space, and at most one stalk S_x has only one element.

Boolean spaces are so special and so useful because they enjoy the *finite partition property* (FPP) : for every open cover $\{U_i \mid i \in I\}$ of a Boolean space X there is a finite set $\{V_1,\dots,V_n\}$ of pairwise disjoint clopen sets such that $X = V_1 \cup \dots \cup V_n$ and $\forall j \le n \; \exists i \in I \;\; V_j \subseteq U_i$.

3.4 LEMMA: Let $\underset{\sim}{S} = (S,\eta,X)$ be a sheaf of algebras over a boolean space X . Then the following hold :

(1) (*patchwork principle*) If $V_1,\dots,V_n$ are pairwise disjoint clopen sets on X and $\sigma_1,\dots,\sigma_n \in \Gamma$ then $\exists \sigma \in \Gamma\underset{\sim}{S} \; \forall i \le n \;\; x \in V_i \Rightarrow \sigma(x) = \sigma_i(x)$.

(2) $\forall s\in S\ \exists\sigma\in\Gamma\underset{\sim}{S}\ \ s = \sigma(x)\ \&\ x = \eta(s)$

(3) $\Gamma\underset{\sim}{S}$ is a subdirect product of the algebras $\underline{S}_x$ $(x\in X)$

(4) S is a Hausdorff-space iff all sets $Eq(\sigma,\tau)$ $\sigma,\tau\in\Gamma\underset{\sim}{S}$ are clopen in X.

(5) If S is Hausdorff and C is a clopen subset of X which contains no $x\in X$ with $|S_x| = 1$, then there exist $\sigma,\tau\in\Gamma\underset{\sim}{S}$ such that $C = Df(\sigma,\tau):=\{x\in X|\sigma x\neq\tau x\}$.

<u>Proof</u>: (1) W.l.o.g. we can assume $V_1\cup\ldots\cup V_n = X$ then it is easy to see that $\sigma_{1|V_1}\cup\sigma_{2|V_2}\cup\ldots\cup\sigma_{n|V_n} =:\sigma$ is a section.

(2): Let τ be an arbitrary section, let N be a neighborhood of s such that ηN is clopen and $\eta:N\longrightarrow\eta N$ is a homeomorphism, and let $\mu:\eta N\longrightarrow N$ be its inverse. Choose $\sigma:=\tau_{|X\setminus\eta N}\cup\mu$, then σ is a section and satisfies (2).

(3): $\Gamma\underset{\sim}{S}$ is obviously a subalgebra of $\Pi\underline{S}_x$, and (2) says that the canonical projections are onto.

(4) First let S be a Hausdorff space, $\sigma,\tau\in\Gamma\underset{\sim}{S}$ and $x\notin Eq(\sigma,\tau)$ i.e. $\sigma(x)\neq\tau(x)$. As S is Hausdorff we can find disjoint neighborhoods N of $\sigma(x)$ and M of $\tau(x)$ such that ηN and ηM are clopen in X and $\eta:N\longrightarrow\eta N$, $\eta:M\longrightarrow\eta M$ are homeomorphisms. The set $U := \sigma^{-1}N\cap\tau^{-1}M$ is an open neighborhood of x and because M and N are disjoint we have $U\cap Eq(\sigma,\tau) = \emptyset$, so we have shown that also the complement of $Eq(\sigma,\tau)$ is open.
To see the opposite direction let $s,t\in S$ and $s\neq t$.
If $\eta s = x\neq y = \eta t$ choose disjoint neighborhoods U of x and V of y. Then $\eta^{-1}U$ and $\eta^{-1}V$ are disjoint neighborhoods of s and t respectively. If $\eta s = x = \eta t$ choose sections σ,τ such that $\sigma x=s$ and $\tau x=t$ (by 2). $Eq(\sigma,\tau)$ is clopen by assumption, so $U := X\setminus Eq(\sigma,\tau)$ is an open neighborhood of x such that σU and τU are disjoint neighborhoods of s and t respectively.

(5): For every $x\in C$ $|S_x|\geq 2$, so we can find $\sigma,\tau\in\Gamma\underset{\sim}{S}$ such that $\sigma x\neq\tau x$ by (2). By (4) there is a whole neighborhood of x in C where σ and τ differ. These neighborhoods cover C, so by FPP we find pairwise disjoint clopen subsets $V_1,\ldots,V_n$ of C and sections $\sigma_1,\ldots,\sigma_n,\tau_1,\ldots,\tau_n\in\Gamma\underset{\sim}{S}$ such that $V_1\cup\ldots\cup V_n = C$ and $x\in V_i \Rightarrow \sigma_i x\neq\tau_i x$. Choose $V_0 := X\setminus C$ and $\sigma_0 = \tau_0 = \sigma_1$ then by (1) we have $\sigma,\tau\in\Gamma\underset{\sim}{S}$ $\forall i\leq n\ x\in V_i \Rightarrow \sigma x = \sigma_i x\ \&\ \tau x = \tau_i x$, hence $C = Df(\sigma,\tau)$. □

If we define the *restriction* $\underset{\sim}{S}|U$ of a sheaf $\underset{\sim}{S}=(S,\eta,X)$ to a subset U of X by $\underset{\sim}{S}|U := (S|U,\eta,U)$, where $S|U := \{s\in S \mid \eta s\in U\}$, then $\underset{\sim}{S}|U$ is again a sheaf of algebras.

3.5 LEMMA: Let $\underset{\sim}{S}= (S,\eta,X)$ be a sheaf of algebras over a Boolean space X and moreover let S be a Hausdorff-space.

(1) If $V_1,\ldots,V_n$ are pairwise disjoint clopen subsets of X such that $X = V_1\cup\ldots\cup V_n$ then $\Gamma\underset{\sim}{S} \simeq \Gamma(\underset{\sim}{S}|V_1)\times\Gamma(\underset{\sim}{S}|V_2)\times\ldots\times\Gamma(\underset{\sim}{S}|V_n)$.

(2) If U is a dense subset of X, then the restriction map $\rho:\Gamma\underset{\sim}{S} \longrightarrow \Gamma(\underset{\sim}{S}|U)$ is an embedding.

(3) $E := \{x\in X \mid |S_x|=1\}$ is a closed subset of X

(4) If $U\subseteq E$ is open, then $\Gamma\underset{\sim}{S} \simeq \Gamma(\underset{\sim}{S}|X\setminus U)$

(5) If E is nowhere dense in X, X_0 is the space X with E contracted to one point x_0 and $\underset{\sim}{S}_0$ the corresponding Boolean sheaf over X_0. Then $\Gamma\underset{\sim}{S} \simeq \Gamma\underset{\sim}{S}_0$.

Proof: (1) is an immediate consequence of 3.4(1).

(2): The restriction map is clearly a homomorphism as the operations are defined componentwise, and it is one-to-one because no two different continuous maps agree on a dense set.

(3): $E = \bigcap\{E(\sigma,\tau) \mid \sigma,\tau\in\Gamma\underset{\sim}{S}\}$ is closed in X as all $E(\sigma,\tau)$ are closed by 3.4(4).

(4): Each continuous section $\sigma:X\setminus U \longrightarrow S|X\setminus U$ obviously has a unique extension $\tau:X \longrightarrow S$ such that $\eta\circ\tau = id_X$.

(5) As the contraction map $\gamma:X \longrightarrow X_0$ contracts only points with one-element stalks, each section $\sigma:X_0 \longrightarrow S_0$ lifts to exactly one section $\tau:X \longrightarrow S$ such that for the contraction map $\delta: S \longrightarrow S_0$ we have $\delta\circ\tau = \sigma\circ\gamma$. □

The previous lemma justifies our restriction to Boolean sheaves. We now want to draw some conclusions from the previous lemmas for Boolean sheaves. We will see that on a Boolean sheaf the topologies both on the base-space and on the sheaf-space are completely determined by the algebra of sections.

3.6 COROLLARY : Let $\underset{\sim}{S} = (S,\eta,X)$ be a Boolean sheaf. Then

(1) If $\underset{\sim}{S}$ has no one-element stalks, the sets $Eq(\sigma,\tau)$ $\sigma,\tau\in\Gamma\underset{\sim}{S}$ form the Boolean algebra X^* of all clopen subsets of X and therefore are a basis for the topology on X .

(2) If $\underset{\sim}{S}$ has a one-element stalk, the sets $Eq(\sigma,\tau)$ and their complements $Df(\sigma,\tau)$ $\sigma,\tau\in\Gamma\underset{\sim}{S}$ form the Boolean algebra X^* of all clopen subsets of X and therefore are a basis for the topology on X .

(3) The topology on S has $\{\sigma U \mid \sigma\in\Gamma\underset{\sim}{S}\ ,\ U\in X^*\}$ as basis.

Proof : (1) and (2) are immediate consequences of 3.4(6) and (3) follows from the fact that η is a local homeomorphism and from 3.4(2). □

We are now going to turn to some model theoretic properties of Boolean sheaves or, to be precise, of the algebras of their sections.

3.7 LEMMA : Suppose $\underset{\sim}{S} = (S,\eta,X)$ is a Boolean sheaf. In the language of $\Gamma\underset{\sim}{S}$ we consider an open formula $\phi(x_1,\dots,x_n)$ and polynomials $p_i(x_1,\dots,x_n)$, $q_i(x_1,\dots,x_n)$ $1\le i\le k$.
Consider the following formulae for $1\le i\le k$:

$$\psi(x_1,\dots,x_n) := \phi(x_1,\dots,x_n)\ \&\ \mathop{\&}_{i=1}^{n} p_i(x_1,\dots,x_n)\neq q_i(x_1,\dots,x_n)$$

$$\psi_i(x_1,\dots,x_n) := \phi(x_1,\dots,x_n)\ \&\ p_i(x_1,\dots,x_n) \neq q_i(x_1,\dots,x_n).$$

Then for $m\le n$ and $\sigma_1,\dots,\sigma_n\in\Gamma\underset{\sim}{S}$ the following hold:

(1) $\{x\in X \mid S_x \models \psi(\sigma_1 x,\dots,\sigma_n x)\}$ is clopen in X .

(2) $\{x\in X \mid S_x \models \exists x_1\dots\exists x_m\ \psi(x_1,\dots,x_m,\sigma_{m+1}x,\dots,\sigma_n x)\}$ is open in X

(3) $\Gamma\underset{\sim}{S} \models \exists x_1\dots\exists x_m\ \phi(x_1,\dots,x_m,\sigma_{m+1},\dots,\sigma_n)$ iff
$\forall x\in X\ S_x \models \exists x_1\dots\exists x_m\ \phi(x_1,\dots,x_m,\sigma_{m+1}x,\dots,\sigma_n x)$

(4) $\Gamma\underset{\sim}{S} \models \exists x_1\dots\exists x_m\ \psi(x_1,\dots,x_m,\sigma_{m+1},\dots,\sigma_n)$ implies
$\forall x\in X\ S_x \models \exists x_1\dots\exists x_m\ \phi(x_1,\dots,x_m,\sigma_{m+1}x,\dots,\sigma_n x)$ and
$\exists x_1\dots\exists x_k\ \ x_i\in X\ \&\ S_{x_i} \models \exists x_1\dots\exists x_m\ \psi_i(x_1\dots x_m,\sigma_{m+1}x_i\dots\sigma_n x_i)$
The opposite implication holds, provided X has no isolated points x with $|S_x|\ge 2$.

Proof: (1): Replace in the open formula ψ each "and" by $\cap$, each "or" by $\cup$, each equation $p(x_1,\dots,x_n) = q(x_1,\dots,x_n)$ by the equalizer $Eq(p(x_1,\dots,x_n),q(x_1,\dots,x_n))$,and each inequality by the complement of the corresponding equalizer. We then get a Boolean expression in X^* and the resulting set is $\{x\in X \mid S_x \models (\sigma_1 x,\dots,\sigma_n x)\}$ and belongs to X^* .

(2) Do the same as in (1), only replace each equation $p(x_1\dots x_n) = q(x_1\dots x_n)$ by the union of the equalizers

$\cup\{Eq(p(\tau_1\dots\tau_m,\sigma_{m+1}\dots\sigma_n),q(\tau_1\dots\tau_m,\sigma_{m+1}\dots\sigma_n)) \mid \tau_1\dots\tau_m \in \Gamma\underset{\sim}{S}\}$ and each inequality by the union over the complements of the corresponding equalizers. The resulting set is the desired set in (2) and it is clearly open.

(3) "=>" is obvious, so we have to check "<=" : For every point $x\in X$ we find a neighborhood U_x and sections $\tau_1^x,\dots,\tau_m^x$ such that $y\in U_x \Rightarrow \phi(\tau_1^x(y)\dots\tau_m^x(y),\sigma_{m+1}(y)\dots\sigma_n(y))$ holds in S_y by (1). FPP guarantees the existence of disjoint clopen sets $V_1\dots V_r$ and sections $\tau_j^i \in \Gamma\underset{\sim}{S}$ $1\le i\le r$, $1\le j\le m$ such that

$X = V_1\cup\dots\cup V_r$, and for each $i\le r$ and $x\in V_i$ S_x satisfies $\phi(\tau_1^i(x),\dots,\tau_m^i(x),\sigma_{m+1}(x),\dots,\sigma_n(x))$. By 3.4(1) we have sections $\tau_1\dots\tau_m$ such that $x\in V_i \Rightarrow \tau_j(x) = \tau_j^i(x)$, $j\le m$.

Thus we have $\Gamma\underset{\sim}{S} \models \phi(\tau_1,\dots,\tau_m,\sigma_{m+1},\dots,\sigma_n)$.

(4) The implication stated here is trivial, but the converse under the additional assumption that X has no isolated points with non-singleton stalks needs a proof. The basic problem, why the same proof as above does not work, is that for two different inequalities $p_i(..)\ne q_i(..)$ and $p_j(..)\ne q_j(..)$ only one stalk S_x might provide solutions but maybe not a common one. But in this case because of (1) x has to be an isolated point with a stalk with at least two elements. So if X has no isolated points with non-singleton stalks, we can assume that there are k <u>different</u> elements $x_1,\dots,x_k\in X$ such that S_{x_i} satisfies

$\exists x_1\dots\exists x_m \psi_i(x_1,\dots,x_m,\sigma_{m+1}(x_i),\dots,\sigma_n(x_i))$. So starting with solutions of ψ_i on a clopen neighborhood of x_i and with solutions for ϕ on the remaining points, the same construction as in the proof of (3) yields sections $\tau_1,\dots,\tau_m\in\Gamma\underset{\sim}{S}$ such that $\Gamma\underset{\sim}{S} \models \psi(\tau_1,\dots,\tau_m,\sigma_{m+1},\dots,\sigma_n)$. □

In most of the applications the sheaf-representation of an algebra is too complicated and we have to find a more special representation of the algebra. The most special case of a sheaf is the so called *constant* (or *trivial* or *simple*)sheaf

$\underset{\sim}{C}(X,\underline{S})$ where X is a topological space and $\underline{S}$ is an algebra. X is the base space and the sheaf space is $S\times X$ where S carries the discrete topology, moreover the local homeomorphism is the second projection $\pi_2 : S\times X \longrightarrow X$. The algebraic structure on the stalk $S\times\{x\}$ is defined in the obvious way such that the first projection $\pi_1 : S\times X \longrightarrow S$ restricted to the stalks becomes an isomorphism. It is clear that the sections of $\underset{\sim}{C}(X,\underline{S})$ are just the continuous maps $X \longrightarrow S$ in the sense that $\sigma : x \longmapsto (s_x, x)$ is a section iff $\sigma : x \longmapsto s_x$ is continuous.

In this context we need the notion of a *subsheaf* of a sheaf which is defined as follows. A sheaf $\underset{\sim}{S}' = (S', \eta', X')$ is called a subsheaf of $\underset{\sim}{S} = (S,\eta,X)$ iff $X' = X$ & $S' \subseteq S$ & $\eta' = \eta|S'$. Because $\underset{\sim}{S}'$ is a sheaf S' has to be an open subset of S and the stalks $\underline{S'}_x$ have to be subalgebras of $\underline{S}_x$ and $\eta S' = X$. Conversely every subset S' of S with these three properties forms a subsheaf of $\underset{\sim}{S}$. The main question we are investigating in the rest of this section is, whether a given sheaf can be embedded as a subsheaf into a constant sheaf.

To make this precise let $\underset{\sim}{S} = (S,\eta,X)$ be a sheaf of algebras and $\underline{A}$ an algebra of the same type. An embedding of $\underset{\sim}{S}$ into $\underset{\sim}{C}(X,\underline{A})$ will be a homeomorphism $f : S \longrightarrow S' \subseteq A\times X$ such that S' is an open subset of $A\times X$, $\eta = \pi_2 \circ f$ and for every $x\in X$ the map $\pi_1 \circ f : S_x \longrightarrow A$ is a homomorphism (which is automatically one-to-one). In this case we write $f : \underset{\sim}{S} \hookrightarrow \underset{\sim}{C}(X,\underline{A})$ and we say $\underset{\sim}{S}$ *is embedded into* $\underset{\sim}{C}(X,\underline{A})$ *as a subsheaf*.
We say f is an *isomorphism* iff f is onto $A\times X$ and in this case we write $\underset{\sim}{S} \cong \underset{\sim}{C}(X,\underline{A})$.

3.8 LEMMA : Let $\underset{\sim}{S} = (S,\eta,X)$ be a Boolean sheaf and let $\underline{A}$ be a finite algebra.

(1) For all $\sigma_a \in \Gamma\underset{\sim}{S}$ $(a\in A)$ the set $\{x\in X \mid a \longmapsto \sigma_a x$ is an embedding $\underline{A} \longrightarrow \underline{S}_x\}$ is clopen in X.

(2) $\{X\in X \mid \underline{A}$ is embeddable into $\underline{S}_x\}$ is open in X.

Proof: There is a first-order open formula $\Psi(x_1,\dots,x_n)$ where $A = \{a_1,\dots,a_n\}$ saying that the mapping $a_i \longmapsto x_i$ $(1\le i\le n)$ is an isomorphism. Now (1) follows from 3.7 and (2) follows from (1), as the set in (2) is a union of those in (1). □

In the sequel we will be dealing with somewhat more special Boolean spaces which are a slight generalization of what Pierce calls *partitionable* Boolean spaces. We say a Boolean space has the *disjoint refinement property* (DRP) if for every collection of open subsets $\{U_i | i \in I\}$ there is a collection $\{V_j | j \in J\}$ of pairwise disjoint open sets such that (i) $\cup\{U_i | i \in I\} = \cup\{V_j | j \in J\}$ and (ii) $\forall j \in J\ \exists i \in I\ V_j \subseteq U_i$

A sheaf $\underset{\sim}{S}$ is said to be *locally embeddable* into $\underset{\sim}{C}(X,\underline{A})$ if for every $x \in X$ there is a neighborhood U of x such that $\underset{\sim}{S}|U$ is embeddable into $\underset{\sim}{C}(U,\underline{A})$ as a subsheaf.

3.9 COROLLARY : Let $\underset{\sim}{S} = (S,\eta,X)$ be a Boolean sheaf and $\underline{A}$ a finite algebra.

(1) If the sets U_i $(i \in I)$ are disjoint open subsets of X, $V := \cup\{U_i | i \in I\}$ and $f_i : \underset{\sim}{S}|U_i \hookrightarrow \underset{\sim}{C}(U_i,\underline{A})$, then $f = \cup\{f_i | i \in I\}$ is an embedding $f : \underset{\sim}{S}|V \hookrightarrow \underset{\sim}{C}(V,\underline{A})$.
If all f_i are isomorphisms then also f is an isomorphism.

(2) If $\sigma_a \in \Gamma\underset{\sim}{S}$ $(a \in A)$ such that for every $x \in U$, U open in X, $a \mapsto \sigma_a x$ is an isomorphism $\underline{A} \to \underline{S}_x$, then $f : S|U \longrightarrow A \times X$ $f(s) := (a,x)$, where $x = \eta s$ and $s = \sigma_a x$, is an isomorphism $f : \underset{\sim}{S}|U \hookrightarrow \underset{\sim}{C}(U,\underline{A})$

(3) If all stalks $\underline{S}_x$ are isomorphic to $\underline{A}$, then $\underset{\sim}{S} \cong \underset{\sim}{C}(X,\underline{A})$.

(4) If S is locally embeddable into $\underset{\sim}{C}(X,\underline{A})$ then S is embeddable into $\underset{\sim}{C}(X,\underline{A})$ as a subsheaf.

(5) If X has the DRP and U is open in X, such that $\underset{\sim}{S}|U$ is locally embeddable into $\underset{\sim}{C}(U,\underline{A})$, then $\underset{\sim}{S}|U$ is embeddable into $\underset{\sim}{C}(U,\underline{A})$ as a subsheaf.

Proof: (1) is obvious and for (2) observe that the sets $\sigma_a(N \cap U)$ with $a \in A$ and N clopen in X form a basis of the topology on $S|U$ by 3.6, hence f is an isomorphism.

(3): Pick for every stalk $\underline{S}_x$ sections σ_a $(a \in A)$ such that $a \mapsto \sigma_a x$ is an isomorphism $\underline{A} \to \underline{S}_x$. By 3.8.(1) the same sections act the same way on a whole neighborhood U of x thus by (2) we have an isomorphism $f_U : \underset{\sim}{S}|U \hookrightarrow \underset{\sim}{C}(U,\underline{A})$. By the FPP of X we can find disjoint clopen sets $V_1,\ldots,V_n$ and isomorphisms $f_i : \underset{\sim}{S}|V_i \hookrightarrow \underset{\sim}{C}(V_i,\underline{A})$ such that $X = V_1 \cup \ldots \cup V_n$. By (1) $f = f_1 \cup \ldots \cup f_n$ is an isomorphism $f : \underset{\sim}{S} \hookrightarrow \underset{\sim}{C}(X,\underline{A})$.

(4) the proof is just the last part of the proof of (3) above.

(5) the proof is again the same if one replaces FPP by DRP and the finite number n by an infinite cardinal. □

For the next lemma, which gives the main construction step for embeddings of arbitrary sheaves into constant sheaves, we need the notion of the *amalgamation of an algebra* $\underline{B}$ *over an algebra* $\underline{A}$, that is an algebra $\underline{C}$ together with an embedding $f:\underline{A} \rightarrowtail \underline{C}$ such that for every embedding $g:\underline{A} \rightarrowtail \underline{B}$ there exists an embedding $\bar{g}:\underline{B} \rightarrowtail \underline{C}$ such that $f = \bar{g} \circ g$.

3.10 LEMMA : Let $\underset{\sim}{S} = (S,\eta,X)$ be a Boolean sheaf, $\underline{A},\underline{B}$ finite algebras and $U \subseteq X$ open such that

(*) $\underline{A}$ is embeddable into each stalk $\underline{S}_x$

(**) $\underset{\sim}{S}|U$ is embeddable into $\underset{\sim}{C}(U,\underline{B})$ as a subsheaf

(***) For $x \notin U$ we have $\underline{S}_x \cong \underline{A}$.

Then for every amalgamation $j:\underline{A} \rightarrowtail \underline{C}$ of $\underline{B}$ over $\underline{A}$, $\underline{C}$ finite $\underset{\sim}{S}$ is embeddable into $\underset{\sim}{C}(X,\underline{C})$ as a subsheaf.

Proof: As in the proof of 3.9(3) we can find sections $\sigma_a \in \Gamma\underset{\sim}{S}$ $(a \in A)$ such that for every $x \in X$ $h_x : a \mapsto \sigma_a x$ is an embedding $h_x:\underline{A} \rightarrowtail \underline{S}_x$. Thus for $x \notin U$ h_x is an isomorphism by (*) and (***). Let $f:\underset{\sim}{S}|U \hookrightarrow \underset{\sim}{C}(U,\underline{B})$ be the embedding in (**). The sections σ_a restricted to U are mapped onto sections $\tau_a \in \underset{\sim}{C}(U,\underline{B})$ by $\tau_a x := f \circ \sigma_a x$ $(x \in U)$ and $f \circ h_x : \underline{A} \longrightarrow \underline{B}$ is an embedding. As for $\zeta \in \Gamma\underset{\sim}{C}(U,\underline{B})$ and $b \in B$ the set $\{x \in U | \zeta(x) = b\}$ is open, we have for every embedding $g:\underline{A} \rightarrowtail \underline{B}$ the open set $V_g := \{x \in U | f \circ h_x = g\}$ and of course the V_g's are pairwise disjoint. For g we have an embedding $\bar{g}:\underline{B} \rightarrowtail \underline{C}$ such that $j = \bar{g} \circ g$, so we have embeddings $\bar{g}_{V_g} : \underset{\sim}{C}(V_g,\underline{B}) \hookrightarrow \underset{\sim}{C}(V_g,\underline{C})$ by setting $\bar{g}_{V_g}(b,x) := (\bar{g}(b),x)$ for $x \in V_g$, $b \in B$. By 3.9(1) these embeddings join to an embedding $\bar{g}_U : \underset{\sim}{C}(U,\underline{B}) \hookrightarrow \underset{\sim}{C}(U,\underline{C})$ and the sections τ_a $(a \in A)$ are all mapped onto the constant sections with value $j(a)$, because $\bar{g}_U \circ \tau_a(x) = \bar{g}_U(\sigma_a(x),x) = \bar{g}_{V_g}(g(a),x) = (j(a),x)$ $x \in V_g$.

Now consider the following mapping $h:S \longrightarrow \underline{C} \times X$ which is defined by

$$h(s) := \begin{cases} (j(a), s) & \text{if } \eta s \notin U \text{ and } \sigma_a(\eta s) = s \\ \bar{g}_U \circ f(s) & \text{if } \eta s \in U. \end{cases}$$

Observe that under h the sections σ_a are mapped to the constant sections $x \mapsto (j(a),x)$ of $\underset{\sim}{C}(X,\underline{C})$ with value $j(a)$. As h coincides with $\bar{g}_U \circ f$ on $S|U$, we only have to check the continuity of h on $S|(X\setminus U)$ in order to show that h is an embedding. For $s \in S|(X\setminus U)$ there is a section σ_a such that $s = \sigma_a(\eta s)$ and $\eta s \notin U$ by (*). Let N be a clopen neighborhood of ηs in X ; then $\{(j(a),x) | x \in N\}$ is a neighborhood of $h(s)$ in $\underline{C} \times X$ and its inverse image under h is the open set $\{\sigma_a(x) | x \in N\}$. □

At the end of this section we try to indicate an answer to the question: "what is so special about subsheaves of a constant sheaf?". The main reason why they are so nice is the fact, that they are already determined by their base space and their stalks, whereas in general the sheaf space could have many different topologies. Also the algebras of sections of these sheaves are very special and in [24] they were called n-Boolean powers. In the special case of a constant sheaf $\underset{\sim}{C}(X,\underline{A})$ the algebra $\underset{\sim}{C}(X,\underline{A})$ is called a *(bounded) Boolean power* of $\underline{A}$ and is denoted by $\underline{A}[B]$ where B is the Boolean algebra of all clopen subsets of X.

3.11 LEMMA : Let X be a Boolean space, $\underline{A}$ a finite algebra, S a subset of $A \times X$, and $A_x := \{a \in A | (a,x) \in S\}$. Then the following are equivalent:

(1) $\underset{\sim}{S} = (S, \pi_2, X)$ is a subsheaf of $\underset{\sim}{C}(X,\underline{A})$

(2) S is open in $A \times X$, $x \in X \Rightarrow A_x \neq \emptyset$ & $A_x \leq \underline{A}$
($\underline{B} \leq \underline{C}$ stands for "$\underline{B}$ is subalgebra of $\underline{A}$")

(3) $\underline{B} \leq \underline{A} \Rightarrow U_{\underline{B}} := \{x \in X | \underline{B} \leq \underline{A}_x \leq \underline{A}\}$ is open in X and
$X = \bigcup\{U_{\underline{B}} | \emptyset \neq \underline{B} \leq \underline{A}\}$

(4) For every subalgebra $\underline{B} \neq \emptyset$ of $\underline{A}$ there exists an open subset $U_{\underline{B}}$ of X such that
$X = \bigcup\{U_{\underline{B}} | \emptyset \neq \underline{B} \leq \underline{A}\}$, $S = \bigcup\{B \times U_{\underline{B}} | \emptyset \neq \underline{B} \leq \underline{A}\}$ and
(*) $U_{\underline{B}} \cap U_{\underline{C}} \neq \emptyset \Rightarrow \exists \underline{D} \leq \underline{A}\ \underline{B},\underline{C} \leq \underline{D}$ & $U_{\underline{B}} \cap U_{\underline{C}} \subseteq U_{\underline{D}}$

Proof : (1)<=>(2) is the definition of a subsheaf. For (2)=>(3) we have to show that $U_{\underline{B}}$ is open. Let $x \in U_{\underline{B}}$, so $\underline{B} \leq A_x$. There exist open neighborhoods N_b of x such that $\{(b,y) | y \in N_b\} \subseteq S$, as S is open in $A \times X$. As B is finite $N = \bigcap\{N_b | b \in B\}$ is open and $x \in N \subseteq U_{\underline{B}}$. (3)=>(4) The sets $U_{\underline{B}}$ defined in (3) obviously satisfy the conditions in (4). (4)=>(2): S is open because all $B \times U_{\underline{B}}$ are open. By condition (*) for every $x \in X$ there is a largest subalgebra $\underline{B}$ of $\underline{A}$ such that $x \in U_{\underline{B}}$, and obviously for this subalgebra we have $A_x = \underline{B}$. □

<u>3.12 COROLLARY :</u> Let X be a Boolean space, $\underline{A}$ a finite algebra, $\underset{\sim}{S} = (S, \pi_2, X)$ a subsheaf of $\underset{\sim}{C}(X, \underline{A})$, and for $\underline{B} \leq \underline{A}$ let $X_{\underline{B}} := \{x \in X \mid A_x \subseteq B\}$.

Then all $X_{\underline{B}}$ $(\underline{B} \leq \underline{A})$ are closed and

$S \cong \{\sigma : X \longrightarrow A \mid \sigma \text{ continuous } \& \ \underline{B} \leq \underline{A} \Rightarrow \sigma(X_{\underline{B}}) \subseteq B\}$.

<u>Proof</u>: $X_{\underline{B}} = \bigcap\{X \setminus U_{\underline{C}} \mid C \nsubseteq B\}$ is closed and $\sigma : X \longrightarrow A$ corresponds to a section $x \longmapsto (\sigma(x), x)$ iff σ is continuous and $x \in X_{\underline{B}}$ implies $\sigma(x) \in B$. □

§4 THE TOPOLOGICAL REPRESENTATION

In this chapter we give a sheaf-representation for all algebras in a discriminator variety extending the results of KEIMEL-WERNER [88]. The proof runs along the lines of [88], but one could also obtain a sheaf-representation with WOLF's [180] representation of arithmetical algebras by sections of sheaves. In Wolf's representation, however, the base space is the space of maximal ideals in the congruence lattice of the algebra, whereas in our case it is just the space of maximal congruences of the algebra in question.

Throughout this section let $\underline{R}$ always be a member of a discriminator variety $\mathcal{V}$. $C(\underline{R})$ denotes the congruence lattice of $\underline{R}$ with the greatest element $\nabla = R \times R$ and the smallest element $\Delta = \{(x,x) \mid x \in R\}$.

We define a topological space $Spec\underline{R}$, called the *spectrum* of $\underline{R}$, which has the set $\{\Phi \in C(\underline{R}) \mid \Theta \in C(\underline{R}) : \Phi \subseteq \Theta \Rightarrow \Phi = \Theta \text{ or } \Theta = \nabla\}$ as underlying set and is endowed with the *equalizer-topology*, i.e. the topology generated by the sets $E(r,s) := \{\Phi \in Spec\underline{R} \mid (r,s) \in \Phi\}$ and their complements $D(r,s) := \{\Phi \in Spec\underline{R} \mid (r,s) \notin \Phi\}$ $r,s \in R$. The $\in$-relation between $R \times R$ and $Spec\underline{R}$ defines a galois-connection

$$V: \mathcal{P}(R\times R) \longrightarrow \mathcal{P}(Spec\underline{R})$$
$$M \longmapsto \{\Phi\in Spec\underline{R} \mid \forall (r,s)\in M\ (r,s)\in\Phi\}$$
$$T: \mathcal{P}(Spec\underline{R}) \longrightarrow \mathcal{P}(R\times R)$$
$$C \longmapsto \{(r,s)\in R\times R \mid \forall\Phi\in C\ (r,s)\in\Phi\}$$

4.1 LEMMA : The sets $E(r,s)$ and $D(r,s)$ $(r,s\in R)$ form a basis of clopen sets for the topology on *Spec*$\underline{R}$, in fact they form a Boolean algebra.

Proof: It suffices to show that the sets $E(r,s)$ and $D(r,s)$ $(r,s\in R)$ are closed under formation of intersections, because together with the fact that they are closed under the formation of complements they form a Boolean algebra which therefore is the basis of a topology consisting of clopen sets only. We are going to show

$E(r,s)\cap E(u,v) = E(t(r,s,u),t(s,r,v))$
$E(r,s)\cap D(u,v) = D(t(r,s,u),t(r,s,v))$
$D(r,s)\cap D(u,v) = D(t(r,t(r,s,u),u),t(r,t(r,s,v),v))$.

First observe that $\nabla\notin D(r,s)$ but $\nabla\in E(r,s)$ for all $r,s\in R$.
Now if Φ is a maximal congruence on $\underline{R}$ we have for $x,y,z\in R$
$(x,y)\in\Phi \Rightarrow (t(x,y,z),z)\in\Phi$ and $(x,y)\notin\Phi \Rightarrow (t(x,y,z),x)\in\Phi$.

Thus: $(r,s)\in\Phi\ \&\ (u,v)\in\Phi$ iff $(t(r,s,u),t(s,r,v))\in\Phi$
$(r,s)\in\Phi\ \&\ (u,v)\notin\Phi$ iff $(t(r,s,u),t(r,s,v))\notin\Phi$
$(r,s)\notin\Phi\ \&\ (u,v)\notin\Phi$ iff $(t(r,t(r,s,u),u),t(r,t(r,s,v),v))$

4.2 COROLLARY: Let $Cl(Spec\underline{R})$ be the set of all closed subsets of *Spec*$\underline{R}$.

(1) If $C\in Cl(Spec\underline{R})$ and $\nabla\in C$, then $C = \bigcap\{E(r,s) \mid C\subseteq E(r,s)\}$
(2) If $C\in Cl(Spec\underline{R})$ and $\nabla\notin C$, then $C = \bigcap\{D(r,s) \mid C\subseteq D(r,s)\}$

Proof: The sets $E(r,s)$ and $D(r,s)$ form a basis of the topology on *Spec*$\underline{R}$, so every open set is a union of such, hence every closed set is an intersection of their complements. If $\nabla\in C$ then of course no $D(r,s)$ contains C and we have (1) . If $\nabla\notin C$ then C is contained in at least one $D(r,s)$ and because the intersection of $D(r,s)$ and any $E(u,v)$ is of the form $D(x,y)$, we get (2) . □

The above corollary also shows under which conditions a set of maximal congruences on $\underline{R}$ yields a subdirect representation of $\underline{R}$.

4.3 COROLLARY : Let $X \subseteq Spec\underline{R}$ be a set of maximal congruences on $\underline{R}$. Then the following conditions are equivalent:

(i) The family $\{\pi_\Phi : \underline{R} \longrightarrow \underline{R}/\Phi \mid \Phi \in X\}$ of canonical projections is a (faithful) subdirect representation of $\underline{R}$.

(ii) $\bigcap\{\Phi \mid \Phi \in X\} = \Delta$

(iii) The set $X \cup \{\nabla\}$ is dense in $Spec\underline{R}$.

Proof: $\bigcap\{\Phi \mid \Phi \in X\} = \Delta$ iff $\forall x,y \in R \ x \neq y \Rightarrow \exists \Phi \in X \ (x,y) \notin \Phi$
iff $\forall x,y \in R \ (\forall \Phi \in X \ (x,y) \in \Phi) \Rightarrow x=y$
iff $\forall x,y \in R \ X \subseteq E(x,y) \Rightarrow x=y$. □

4.4 LEMMA : For the galois-connection V,T between $R \times R$ and $Spec\underline{R}$ we have:

(1) $M \subseteq R \times R \Rightarrow VM = \bigcap\{E(r,s) \mid (r,s) \in M\} = \{\Phi \in Spec\underline{R} \mid M \subseteq \Phi\}$

(2) $C \in Spec\underline{R} \Rightarrow TC = \bigcap\{\Phi \mid \Phi \in C\}$

(3) $TVM = M \iff M \in C(\underline{R})$

(4) $VTC = C \iff \nabla \in C \in Cl(Spec\underline{R})$

Proof: (1) and (2) follow immediately from the definition and they show that VM is always closed and contains ∇ and that TC is always a congruence. Now for every congruence $\Theta \in C(\underline{R})$ we have by 2.2(4) $\Theta = \bigcap\{\Phi \in Spec\underline{R} \mid \Theta \subseteq \Phi\} = TV\Theta$. If $\nabla \in C \in Cl(Spec\underline{R})$ we have by 4.2(1) $C = \bigcap\{E(r,s) \mid C \subseteq E(r,s)\} = TM$, where $M = \{(r,s) \mid C \subseteq E(r,s)\}$, thus $TVC = TVTM = TM = C$. □

4.5 COROLLARY : The mapping $U : \Theta \longmapsto \{\Phi \in Spec\underline{R} \mid \Theta \nsubseteq \Phi\}$ is a lattice isomorphism between the congruence lattice $C(\underline{R})$ and the lattice of open sets of $Spec\underline{R}$ which don't contain ∇ . Moreover $U(\theta(x,y)) = D(x,y)$.

4.6 THEOREM : For every algebra $\underline{R}$ in a discriminator variety $\mathcal{W}$ the topological space $Spec\underline{R}$ is a Boolean space and its dual $(Spec\underline{R})^*$, the Boolean algebra of all clopen subsets of $Spec\underline{R}$, is just the set $\{E(r,s) \mid r,s \in R\} \cup \{D(r,s) \mid r,s \in R\}$.

Proof: $Spec\underline{R}$ has a basis of clopen sets, so it is 0-dimensional and it remains to show that $Spec\underline{R}$ is a compact Hausdorff space. If $\Phi,\Psi \in Spec\underline{R}$ and $\Phi \nsubseteq \Psi$ then there exist $u,v \in R$ such that $(u,v) \in \Phi$ but $(u,v) \notin \Psi$, thus $E(u,v)$ and $D(u,v)$ are disjoint

neighborhoods of Φ and Ψ respectively. This shows that *Spec*$\underline{R}$ is a Hausdorff space.
Now let $\{U_i \mid i \in I\}$ be an open cover of *Spec*$\underline{R}$. There is an $i_0 \in I$ such that $\nabla \in U_{i_0}$ and moreover there exist $r,s \in R$ such that $E(r,s) \subseteq U_{i_0}$ by 4.1. Define $V_i := U_i \cap D(r,s)$ $(i \in I)$; then $\{V_i \mid i \in I\}$ is an open cover of $D(r,s)$. By 4.4 there are congruences $\Phi_i \in C(\underline{R})$ such that $U\Phi_i = V_i$, and moreover $U(\Theta(r,s)) = D(r,s) = \bigcup\{V_i \mid i \in I\} = \bigcup\{U\Phi_i \mid i \in I\} = U(\bigvee\{\Phi_i \mid i \in I\})$ as U is a lattice isomorphism. By the injectivity of U we get $\Theta(r,s) = \bigvee\{\Phi_i \mid i \in I\}$, and by compactness of $\Theta(r,s)$ we find $i_1,\dots,i_n \in I$ such that $\Theta(r,s) = \Phi_{i_1} \vee \dots \vee \Phi_{i_n}$, but this implies $D(r,s) = V_{i_1} \cup \dots \cup V_{i_n}$ and therefore *Spec*$\underline{R} = U_{i_0} \cup U_{i_1} \cup \dots \cup U_{i_n}$. □

*Spec*__ can be extended in a natural way to a functor in the sense of 4.8 below. In order to establish this result we define for every homomorphism $f: \underline{R} \longrightarrow \underline{S}$ in a discriminator variety $\mathfrak{V}$ and for every $\Phi \in$ *Spec*$\underline{S}$:

Spec$f(\Phi) := f^+(\Phi) := \{(x,y) \mid x,y \in R \ \&\ (fx,fy) \in \Phi\}$.

We are going to show that f^+ is a continuous mapping f^+: *Spec*$\underline{S} \longrightarrow$ *Spec*$\underline{R}$ whose image is the closed subspace $V(\mathit{Ker}f) = \{\Phi \in \mathit{Spec}\underline{R} \mid \mathit{Ker}f \subseteq \Phi\}$ of *Spec*$\underline{R}$, where *Kerf* denotes the congruence $\mathit{Ker}f := \{(x,y) \mid x,y \in R \ \&\ fx=fy\}$ on $\underline{R}$, called the ***kernel*** of f. f^+ also preserves some distinguished subsets $X_{\mathfrak{A}}$ of the spectrum, where $X_{\mathfrak{A}} \subseteq$ *Spec*$\underline{R}$ is defined for every class $\mathfrak{A}$ of simple algebras in $\mathfrak{V}$ as follows:

$X_{\mathfrak{A}} := \{\Phi \in \mathit{Spec}\underline{R} \mid \underline{R}/\Phi$ embeddable into some member of $\mathfrak{A}\}$.

If there is danger of confusion we write $X^{\underline{R}}_{\mathfrak{A}}$, $X^{\underline{S}}_{\mathfrak{A}}$, ... instead of $X_{\mathfrak{A}}$.

<u>4.7 LEMMA</u>: If $\mathfrak{A}$ is a class of simple algebras in $\mathfrak{V}$ which is closed under ultraproducts - in particular if $\mathfrak{A}$ is a finite set of finite simple $\mathfrak{V}$-algebras - then $X_{\mathfrak{A}}$ is a closed subset of *Spec*$\underline{R}$.

<u>Proof</u>: In the subvariety $\mathfrak{U}$ of $\mathfrak{V}$ generated by $\mathfrak{A}$ every simple algebra is embeddable into a member of $\mathfrak{A}$, so we have for $\Theta \in$ *Spec*$\underline{R}$: $\Theta \in X_{\mathfrak{A}} \iff \underline{R}/\Theta \in \mathfrak{U} \iff \Theta \in V\Phi$, where Φ is the smallest congruence on $\underline{R}$ such that $\underline{R}/\Phi \in \mathfrak{U}$. A Φ with that property exists because $\mathfrak{U}$ is closed under subdirect products, hence the family of Φ's such that $\underline{R}/\Phi \in \mathfrak{U}$ is closed under arbitrary intersections. □

4.8 PROPOSITION : Let $\mathcal{V}$ be a discriminator variety, $\underline{R},\underline{S},\underline{T}\in\mathcal{V}$ and $f:\underline{R}\longrightarrow\underline{S}$, $g:\underline{S}\longrightarrow\underline{T}$ homomorphisms.

Then the following hold:

(i) $f^+ : Spec\underline{S} \longrightarrow Spec\underline{R}$

(ii) $(f^+)^{-1}(E(x,y)) = E(fx,fy)$ for $x,y\in R$

(iii) f^+ is continuous

(iv) $f^+(X^{\underline{S}}_{\mathfrak{a}}) \subseteq X^{\underline{R}}_{\mathfrak{a}}$

(v) $f^+(Spec\underline{S}) = V(Ker f)$

(vi) $(g\circ f)^+ = f^+\circ g^+$ and $(id_{\underline{R}})^+ = id_{Spec\underline{R}}$

(vii) If f is onto , then

$\underline{R}/f^+\Phi \cong \underline{S}/\Phi$ and $(f^+)^{-1}(X^{\underline{R}}_{\mathfrak{a}}) = X^{\underline{S}}_{\mathfrak{a}}$

Proof: For $\Phi\in Spec\underline{S}$ $f^+\Phi$ is the inverse image of Φ under f and therefore the algebra $\underline{R}/f^+\Phi$ is embeddable into $\underline{S}/\Phi$. (i) follows because subalgebras of simple algebras are simple, so $f^+\Phi$ is maximal. (iv) also follows from this observation.

(ii): $\Phi\in(f^+)^{-1}(E(x,y)) \iff f^+\Phi\in E(x,y)$
$\iff (x,y)\in f^+\Phi$
$\iff (fx,fy)\in\Phi$.

(iii) is an immediate consequence of (ii) .

(v) is true because in general $Ker f\subseteq\Phi \iff \Phi=f^+(f\Phi)$ holds.

(vi) follows immediately from the definition.

(vii): If f is onto the canonocal embedding $\underline{R}/f^+\Phi \rightarrowtail \underline{S}/\Phi$ is necessarily onto, so it is an isomorphism.

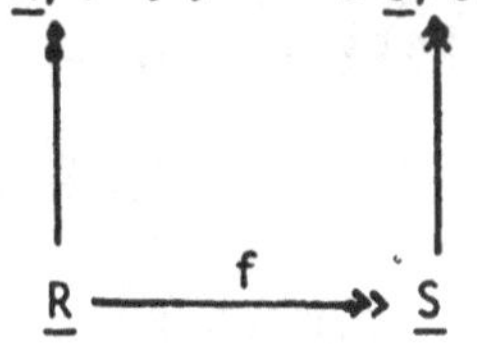

$\Phi\in(f^+)^{-1}(X^{\underline{R}}_{\mathfrak{a}}) \iff f^+\Phi\in X^{\underline{R}}_{\mathfrak{a}} \iff \underline{R}/f^+\Phi$ embeddable into some $\underline{A}\in\mathfrak{A}$
$\iff \underline{S}/\Phi$ embeddable into some $\underline{A}\in\mathfrak{A}$
$\iff \Phi\in X^{\underline{S}}_{\mathfrak{a}}$. □

Collecting the facts from the beginning of this section we have for every algebra $\underline{R}\in\mathcal{V}$ a subdirect representation $\{p_\Phi:\underline{R}\longrightarrow\underline{R}/\Phi \mid \Phi\in Spec\underline{R}\}$ by simple algebras $\underline{R}/\Phi$. We have a natural topology on $Spec\underline{R}$ which turns $Spec\underline{R}$ into a Boolean space and looking into the standard construction 3.3 we see that $\{\Phi \mid p_\Phi(x)=p_\Phi(y)\} = E(x,y)$ are open sets in $Spec\underline{R}$. So the standard construction yields the sheaf $\underset{\sim}{S}(\underline{R}) = (\bigcup\{\underline{R}/\Phi \mid \Phi\in Spec\underline{R}\},\eta,Spec\underline{R})$ over the Boolean space $Spec\underline{R}$, which we call the *standard sheaf associated with* $\underline{R}$. We also get the standard representation $\hat{}:\underline{R}\longrightarrow\Gamma\underset{\sim}{S}(\underline{R})$ of $\underline{R}$.

The main result in this section is

4.9 THEOREM : Let $\mathcal{V}$ be a discriminator variety, $\underline{R}\in\mathcal{V}$, and $\underset{\sim}{S}(\underline{R})$ the standard sheaf associated with $\underline{R}$.
The standard representation $\hat{}: \underline{R} \longrightarrow \Gamma\underset{\sim}{S}(\underline{R})$ of $\underline{R}$ is an isomorphism.

Proof: Let $\sigma\in\Gamma\underset{\sim}{S}(\underline{R})$. We have to show $\exists r\in R\ \sigma=\hat{r}$. As a first step we show that this is locally true and then we use some kind of patchwork principle similarly to 3.4(1) but only for the sections of the form $\hat{r}$ to get the result.
For every $\Phi\in Spec\underline{R}$ we have $\sigma\Phi\in\underline{R}/\Phi$, so $\exists r\in R\ \sigma\Phi=[r]\Phi=\hat{r}\Phi$.
σ and $\hat{r}$ are both sections and agree on Φ, so by 3.2(1) there is a neighborhood U of Φ such that $\Psi\in U \Rightarrow \sigma\Psi=\hat{r}\Psi$.
By the FPP of $Spec\underline{R}$ we get finitely many pairwise disjoint clopen sets $V_0,\dots,V_n$ and elements $r_0,\dots,r_n\in R$ such that $Spec\underline{R} = V_0\cup\dots\cup V_n$ and for $i=0,\dots,n$ $\Phi\in V_i \Rightarrow \sigma\Phi=\hat{r}_i\Phi$. We can assume $\nabla\in V_0$ and $\nabla\notin V_i$ $(i=1,\dots,n)$. Because the sets V_i are clopen, there are elements $s_1,\dots,s_n,t_1,\dots,t_n\in R$ such that $V_i = D(s_i,t_i)$ $(i=1,\dots,n)$. Now define $u_0,\dots,u_n$ recursively as follows: $u_0:=r_0$, $u_{i+1}:=n(s_{i+1},t_{i+1},u_i,r_{i+1})$ $0\le i\le n-1$.
It is clear that all u_i are in $\underline{R}$ and it is easy to see by induction that for $i\le n$ and $\Phi\in\bigcup\{V_j \mid j\le i\}$ we have $\sigma\Phi=\hat{u}_i\Phi$; thus we have shown $\sigma = \hat{u}_n$. □

Making the easy observation that $Eq(\hat{x},\hat{y}) = E(x,y)$ are clopen subsets of $Spec\underline{R}$ we get :

4.10 COROLLARY : Let $\mathcal{V}$ be a discriminator variety, $\underline{R}\in\mathcal{V}$, then the standard sheaf $\underset{\sim}{S}(\underline{R})$ associated with $\underline{R}$ is a Boolean sheaf. (see 3.4(4))

We could have obtained a similar sheaf representation of $\underline{R}\in\mathcal{V}$ by considering the *proper spectrum* $Spec_o\underline{R} := Spec\underline{R}\setminus\{\nabla\}$ of $\underline{R}$. $Spec_o\underline{R}$ of course in general is not a Boolean space - this is only the case if ∇ is an isolated point in $Spec\underline{R}$, which is equivalent to saying that ∇ is a compact congruence.
The same standard sheaf construction works and we get the standard sheaf $\underset{\sim}{S}_0(\underline{R})$ associated with $\underline{R}$ over the proper spectrum of $\underline{R}$, and the standard embedding $\hat{}:\underline{R} \longrightarrow \Gamma\underset{\sim}{S}_0(\underline{R})$.
In general the standard embedding is not onto, namely two sections $\hat{r}$ and $\hat{s}$ differ only on a compact-open subset of the locally Boolean space $Spec_o\underline{R}$ and indeed every section $\sigma\in\Gamma\underset{\sim}{S}_0(\underline{R})$

which differs from some $\hat{F}$ only on a compact-open set, is of the form $\sigma = \hat{s}$, $s \in R$. So $\underline{R}$ has a representation as the algebra $\Gamma_c \underset{\sim}{S}_o(\underline{R})$ of all sections with *compact support* . In the sequel we are always using the representation in 4.9 , so we won't go into more detail here.

At this point we have to say something about the expressive power of the discriminator. In 1.10 we have seen that each open Horn-formula is equivalent to an equation in the presence of the discriminator. Now we are going to show that for each algebra $\underline{R}$ in a discriminator variety $\mathcal{W}$ every first order sentence about the topology on *Spec*$\underline{R}$ can be translated with help of the discriminator into a first order sentence about $\underline{R}$.

For every algebra $\underline{R} \in \mathcal{W}$ let $\mathcal{D}(\underline{R})$ denote the set $\{D(r,s) | r,s \in R\}$ of all clopen sets in *Spec*$\underline{R}$, which don't contain ∇ , in other words $\mathcal{D}(\underline{R})$ is the set of all compact-open subsets of the proper spectrum of $\underline{R}$. Under the galois-correspondence (V,T) above $\mathcal{D}(\underline{R})$ is isomorphic to the lattice of all principal congruences of $\underline{R}$. By lemma 4.1 $\mathcal{D}(\underline{R})$ is a relatively complemented distributive lattice and the operations can be expressed with help of the discriminator as follows:

$D(r,s) \cup D(u,v) = D(t(r,s,u),t(s,r,v))$

$D(r,s) \cap D(u,v) = D(t(r,t(r,s,u),u),t(r,t(r,s,v),v))$

$D(u,v) \setminus D(r,s) = D(t(r,s,u),t(r,s,v))$.

Moreover if $o \in R$ is an arbitrarily chosen element in R , then by 1.14(g) we have $D(u,v) = D(o,s(u,v,o))$.

As an immediate consequence we get

<u>4.11 LEMMA :</u> Let $\mathcal{W}$ be a discriminator variety and $\underline{R} \in \mathcal{W}$.

Pick $o \in R$ and define for $x,y \in R$

$x \wedge y := t(o,t(o,x,y),y)$

$x \vee y := t(x,o,y)$

$x \setminus y := t(o,y,x)$.

Then the mapping $d: R \longrightarrow \mathcal{D}(\underline{R})$, $d(r) := D(o,r)$ is an onto-$(\wedge,\vee,\setminus)$-homomorphism.

This lemma allows us to express by a first-order sentence in the language of $\underline{R}$ whether *Spec*$\underline{R}$ has isolated points etc. .

In the sequel we often have to deal with subdirect retracts of simple algebras in $\mathcal{W}$. $\underline{R} \in \mathcal{W}$ is called a *subdirect retract* of the (simple) algebras $\underline{A}_i$ $(i \in I)$ in $\mathcal{W}$ iff there is an embedding

$f:\underline{R} \rightarrowtail \Pi\{\underline{A}_i | i\in I\}$ and a retraction $g:\Pi\{\underline{A}_i | i\in I\} \twoheadrightarrow \underline{R}$ such that $g\circ f = id_R$ and for each projection $p_i:\Pi\{\underline{A}_i | i\in I\} \longrightarrow \underline{A}_i$ the composit $p_i\circ f:\underline{R} \longrightarrow \underline{A}_i$ is an onto-homomorphism.

If $\underline{A}$ is a finite algebra and $\underline{B}$ is a Boolean algebra, we denote by $\underline{A}[\underline{B}]$ the algebra of all continuous mappings from the Stone-space B^* of $\underline{B}$ into the discrete algebra $\underline{A}$. $\underline{A}[\underline{B}]$ is called a *Boolean power* of $\underline{A}$ (see e.g. BURRIS [22]) and it is isomorphic to the algebra $\Gamma\underset{\sim}{C}(B^*,\underline{A})$ of all sections of the constant sheaf $\underset{\sim}{C}(B^*,\underline{A})$.

4.12 LEMMA : Let $\underline{R}$ be a subdirect retract of the simple algebras $\underline{A}_i\in\mathcal{W}$ $(i\in I)$, $|A_i|\neq 1$.

(1) If $\mathfrak{A}$ is an elementary class of simple algebras in $\mathcal{W}$, then the set $X_{\mathfrak{A}} \subseteq Spec\underline{R}$ is clopen in $Spec\underline{R}$; in particular ∇ is an isolated point in $Spec\underline{R}$.

(2) $\mathcal{D}(\underline{R})$ is a complete Boolean algebra, in fact a retract of $\mathfrak{2}^I$.

(3) If all $\underline{A}_i$ are isomorphic to the finite algebra $\underline{A}$, then $\underline{R} \simeq \underline{A}[\mathcal{D}(\underline{R})]$.

Proof: (1): Let $\underline{S} = \Pi\{\underline{A}_i | i\in I\}$, $f:\underline{R} \rightarrowtail \underline{S}$, $g:\underline{S} \twoheadrightarrow \underline{R}$, and $p_i:\underline{S} \twoheadrightarrow \underline{A}_i$ as above. Observe that $Spec\underline{S}\setminus\{\nabla\}$ is precisely the space of ultrafilters on I and its Stone-dual is the Boolean algebra $\mathfrak{2}^I$.

Consider the mappings $f^+:Spec\underline{S} \longrightarrow Spec\underline{R}$ and $g^+:Spec\underline{R} \longrightarrow Spec\underline{S}$. By 4.8(iv) we have $f^+\circ g^+ = (g\circ f)^+ = (id_R)^+ = id_{Spec\underline{R}}$, in particular f^+ is onto and g^+ is one-to-one.

As g is onto we use 4.8(vii) to get $X^{\underline{R}}_{\mathfrak{A}} = (g^+)^{-1}(X^{\underline{S}}_{\mathfrak{A}})$.

A proper maximal congruence on $\underline{S}$ is just the filter-congruence for an ultrafilter u on I, and $\underline{S}/u$ is embeddable into a member of $\mathfrak{A}$ iff $I_{\mathfrak{A}}:=\{i\in I | \underline{A}_i$ embeddable into a member of $\mathfrak{A}\}$ belongs to u. This says that $X^{\underline{S}}_{\mathfrak{A}}$ is clopen in $Spec\underline{S}$.

Now $X^{\underline{R}}_{\mathfrak{A}}$ as the counterimage of the clopen set $X^{\underline{S}}_{\mathfrak{A}}$ under the continuous map g^+ is a clopen subset of $Spec\underline{R}$.

(2) : The clopen subsets of $Spec\underline{R}$ are precisely those of the form $(g^+)^{-1}(N)$, where N is a clopen subset of $Spec\underline{S}$, moreover $(g^+)^{-1}(\{\nabla_S\})=\{\nabla_R\}$ and $\mathcal{D}(\underline{S}) \simeq \mathfrak{2}^I$. Thus the mappings $f^*:\mathcal{D}(\underline{R}) \longrightarrow \mathcal{D}(\underline{S})$ and $g^*:\mathcal{D}(\underline{S}) \longrightarrow \mathcal{D}(\underline{R})$, where $f^*:=(f^+)^{-1}$ and

$g^*:=(g^+)^{-1}$ show that $\mathcal{D}(\underline{R})$ is a retract of $\mathcal{D}(\underline{S})\cong 2^I$, because $g^*\circ f^* = id_{\mathcal{D}(\underline{R})}$.

(3) is an immediate consequence of 3.10 in the special case $U = \emptyset$ and $\underline{A} \cong \underline{B} \cong \underline{C}$. □

§ 5 SPECIAL EMBEDDINGS

In this section we investigate special embeddings in discriminator varieties, such as essential, pure (=algebraic), existential, and elementary embeddings because they give some information about special algebras in the variety, such as injective, equationally compact., algebraically closed, or existentially closed algebras. Our main tool for the investigation of these embeddings will be the sheaf representation of the algebra in question.

Every embedding $f:\underline{R} \rightarrowtail \underline{S}$ induces an embedding f^* of the Boolean algebras $f^*: Spec\underline{R}^* \rightarrowtail Spec\underline{S}^*$, and for every $\Phi \in Spec\underline{S}$ an embedding $f_\Phi:\underline{R}/f^{-1}\Phi \rightarrowtail \underline{S}/\Phi$; the objective will be to reduce the characterization of special embeddings in the variety to the characterization of special embeddings for Boolean algebras and for the simple members of the variety.

We first give the definitions for some special embeddings, so let $\underline{R}$ be a subalgebra of $\underline{S}$. $\underline{S}$ is called an *essential extension* of $\underline{R}$ and $\underline{R}$ is called *large* in $\underline{S}$, iff the only congruence on $\underline{S}$, which has the identity on $\underline{R}$ as its restriction to $\underline{R}$, is the identity on $\underline{S}$, in other words a homomorphism $f:\underline{S} \longrightarrow \underline{T}$ is one-to-one iff its restriction to $\underline{R}$ $f|\underline{R} : \underline{R} \longrightarrow \underline{T}$ is one-to-one. $\underline{S}$ is said to be an *α-elementary* extension of $\underline{R}$ where α is a first order sentence in the language of $\underline{S}$ with constants from $\underline{R}$, iff $\underline{S}\models\alpha \iff \underline{R}\models\alpha$.

An *embedding* (= one-to-one homomorphism) $f:\underline{R} \rightarrowtail \underline{S}$ is called

essential iff $\underline{S}$ is an essential extension of $f(\underline{R})$.

pure (or *algebraic*) iff $\underline{S}$ is an α-elementary extension of $f(\underline{R})$ for every existentially quantified conjunction α of atomic formulas - atomic formulas are equations with constants from $f(\underline{R})$.

existential iff $\underline{S}$ is an α-elementary extension of $f(\underline{R})$ for every existential sentence α with constants from $f(\underline{R})$.

elementary iff $\underline{S}$ is an α-elementary extension of $f(\underline{R})$ for every sentence α.

In the sequel $\mathcal{W}$ will always be a discriminator variety and $\mathfrak{A}$ its class of simple algebras - the 1-element algebras included. For a homomorphism $f:\underline{R} \longrightarrow \underline{S}$ we have the continuous map $f^{+}:Spec\underline{S} \longrightarrow Spec\underline{R}$ and a Boolean homomorphism $f^{*}:Spec\underline{R}^{*} \longrightarrow Spec\underline{S}^{*}$ $f^{*}:D(r,s) \longmapsto D(fr,fs)$ $r,s \in R$. For every $\Phi \in Spec\underline{S}$ we have the canonical embedding $f_{\Phi}:\underline{R}/f^{-1}\Phi \rightarrowtail \underline{S}/\Phi$ in $\mathfrak{A}$, which is an isomorphism provided f is onto (see 3.8(vii)).

Observe that $Spec\underline{R}$ is a *pointed* Boolean space - the distinguished point in $Spec\underline{R}$ is ∇ - and that f^{+} preserves the distinguished point - that is $f^{+}\nabla_{\underline{S}} = \nabla_{\underline{R}}$.

Correspondingly $Spec\underline{R}^{*}$ is a Boolean algebra with a distinguished filter $\delta_{\underline{R}} := \{E(x,y) | x,y \in R\}$ and again f^{*} preserves these filters, meaning $(f^{*})^{-1}(\delta_{\underline{S}}) = \delta_{\underline{R}}$.

We could avoid this by considering the locally Boolean space $Spec_{o}\underline{R}$ and its "dual" $Spec_{o}\underline{R}^{*} := \{D(x,y) | x,y \in R\}$ which is the relatively complemented lattice of all compact-open subsets of $Spec_{o}\underline{R}$. Unfortunately the restriction of f^{+} to $Spec_{o}\underline{S}$ in general is not a mapping into $Spec_{o}\underline{R}$, because $f^{+}\Phi$ may be ∇ if $\underline{S}/\Phi$ has a 1-element subalgebra, but the restriction of f^{*} to $Spec_{o}\underline{R}^{*}$ is a homomorphism between the relatively complemented lattices $Spec_{o}\underline{R}^{*}$ and $Spec_{o}\underline{S}^{*}$.

We now turn to the characterization of essential embeddings. As in $\mathfrak{A}$ all embeddings are essential except those of a 1-element algebra into a bigger one; it is not very surprising that the characterization is just in terms of the spectra.

<u>5.1 LEMMA :</u> For an embedding $f:\underline{R} \rightarrowtail \underline{S}$ in $\mathcal{W}$ t.f.a.e.

(i) $f:\underline{R} \rightarrowtail \underline{S}$ is essential in $\mathcal{W}$

(ii) A subset C of $Spec\underline{S}$ containing ∇ is dense in $Spec\underline{S}$ iff $f^{+}C$ is dense in $Spec\underline{R}$.

(iii) If C is a closed subset of $Spec\underline{S}$, then $Spec_{o}\underline{R} \subseteq f^{+}C \iff Spec_{o}\underline{S} \subseteq C$

(iv) $f^{*}:Spec_{o}\underline{R}^{*} \rightarrowtail Spec_{o}\underline{S}^{*}$ is essential in the class $\mathcal{R}$ of all relatively complemented distributive lattices.

Proof: For every subset $C \subseteq Spec\underline{S}$ and $\Theta = \bigcap\{\Phi \mid \Phi \in C\}$ we have $f^+\Theta = \bigcap\{f^+\Phi \mid \Phi \in C\} = \bigcap\{\Psi \mid \Psi \in f^+C\}$. By 3.3 we get $C \cup \{\nabla\}$ dense in $Spec\underline{S}$ iff $\bigcap\{\Phi \mid \Phi \in C\} = \Delta$ iff $\Theta = \Delta$. $f^+C \cup \{\nabla\}$ is dense in $Spec\underline{R}$ iff $\bigcap\{\Psi \mid \Psi \in f^+C\} = \Delta$ iff $f^+\Theta = \Delta$.
These remarks immediately show (i)=>(ii) and (iii)=>(i) ; (ii)=>(iii) is trivial and (iii)<=>(iv) follows from the fact that congruences on $Spec_0\underline{R}^*$ uniquely correspond to ideals on $Spec_0\underline{R}^*$, and these correspond uniquely to open subsets of $Spec\underline{R}$ which don't contain ∇. □

The characterization of the other special embeddings seems to be very difficult, and we can only give some sufficient conditions for embeddings to be pure or existential.

5.2 LEMMA : Let $f:\underline{R} \rightarrowtail \underline{S}$ be an embedding in the discriminator variety $\mathcal{W}$. If there is a subset $X \subseteq Spec\underline{S}$ such that $f^+X = Spec\underline{R}\setminus\{\nabla\}$ and for each $\Phi \in X$ $f_\Phi : \underline{R}/f^+\Phi \rightarrowtail \underline{S}/\Phi$ is pure, then $f:\underline{R} \rightarrowtail \underline{S}$ is pure.

Proof: Let $\alpha(x_1,\dots,x_n)$ be an existential positive formula and $a_1,\dots,a_n \in R$.

$$\begin{aligned}
\underline{S} \models \alpha(fr_1,\dots,fr_n) &\Rightarrow \forall \Phi \in Spec\underline{S}\ \ \underline{S}/\Phi \models \alpha([fr_1]\Phi,\dots,[fr_n]\Phi) \\
&\quad \text{as } \alpha \text{ is existential positive} \\
&\Rightarrow \forall \Phi \in X\ \ \underline{R}/f^+\Phi \models \alpha([r_1]f^+\Phi,\dots,[r_n]f^+\Phi) \\
&\quad \text{as } f_\Phi \text{ is pure} \\
&\Rightarrow \forall \Psi \in Spec\underline{R}\ \ \underline{R}/\Psi \models \alpha([r_1]\Psi,\dots,[r_n]\Psi) \\
&\quad \text{as } f^+X = Spec\underline{R}\setminus\{\nabla\} \\
&\Rightarrow \underline{R} \models \alpha(r_1,\dots,r_n) \text{ by 4.10 and 3.7 . } \square
\end{aligned}$$

This lemma generalizes the following result of W.TAYLOR : "Pure compactifications in quasi-primal varieties" Can.J.Math. (to appear) .

5.3 COROLLARY : For every algebra $\underline{R}$ in a discriminator variety the canonical embedding $\underline{R} \rightarrowtail \Pi\{\underline{R}/\Phi \mid \Phi \in Spec\underline{R}\}$ is pure.

Proof : Choose for X the set of all principal ultrafilter congruences on $\Pi\{\underline{R}/\Phi \mid \Phi \in Spec\underline{R}\}$. Then X satisfies the assertions of 5.2 because all f_Φ in question are isomorphisms , hence pure. □

5.4 LEMMA : Let $f:\underline{R} \rightarrowtail \underline{S}$ be an embedding in the discriminator variety $\mathcal{V}$, and let $Spec\underline{R}$ have no isolated points different from ∇. If there is a dense subset X in $Spec\underline{S}\setminus\{\nabla\}$ such that $f^{+}X = Spec\underline{R}\setminus\{\nabla\}$ and for each $\Phi\in X$ $f_{\Phi}:\underline{R}/f^{+}\Phi \rightarrowtail \underline{S}/\Phi$ is existential, then f is existential.

Proof: Let $\psi, p_i, q_i, \phi, \psi_i$ $(i=1,..,k)$ be as in lemma 3.7 , and $r_{m+1},...,r_n \in R$.

$\underline{S} \models \exists x_1 ... \exists x_m\ \psi(x_1,...,x_m, fr_{m+1},...,fr_n)$

$\Rightarrow \quad \forall\Phi\in Spec\underline{S}\quad \underline{S}/\Phi \models \exists x_1 ... \exists x_m\ \phi(x_1,...,x_m,[fr_{m+1}]\Phi,...,[fr_n]\Phi)$

$\quad \& \ \exists\Phi_1,..,\Phi_k \in Spec\underline{S}$ such that for $i=1,...,k$

$\quad\quad \underline{S}/\Phi_i \models \exists x_1 ... \exists x_m\ \psi_i(x_1,...,x_m,[fr_{m+1}]\Phi_i,...,[fr_n]\Phi_i)$

X is dense in $Spec\underline{S}\setminus\{\nabla\}$, all $\Phi_1,...,\Phi_k$ are different from ∇ , and an existential sentence holds in one stalk iff it holds in all stalks in an open neighborhood (3.7(2)), so we can assume $\Phi_1,...,\Phi_k\in X$. As all f_{Φ} are existential, we get

$\forall\Phi\in X\quad \underline{R}/f^{+}\Phi \models \exists x_1 ... \exists x_m\ \phi(x_1,...,x_m,[r_{m+1}]f^{+}\Phi,...,[r_n]f^{+}\Phi)$ and

$\exists\Phi_1,...,\Phi_k\in X\quad R/f^{+}\Phi_i \models \exists x_1 ... \exists x_m\ \psi_i(x_1,...,x_m,[r_{m+1}]f^{+}\Phi_i,...,[r_n]f^{+}\Phi_i)$

As $f^{+}X = Spec\underline{R}\setminus\{\nabla\}$ and $Spec\underline{R}$ has no isolated points different from ∇, we can apply 3.7(4) to get

$\underline{R} \models \exists x_1 ... \exists x_m\ \psi(x_1,...,x_m,r_{m+1},...,r_n)$. □

Let us remark at this point that the property of $Spec\underline{R}$ not to have isolated points can be expressed by an $\forall\exists$-sentence for $\underline{R}$. To get these sentences we just have to use $\Theta(x,z)\subseteq\Theta(x,y)$ iff $t(x,y,z) = x$ in order to state that the lattice of principal congruences on $\underline{R}$ has no atoms and no greatest element, the isolated points of $Spec\underline{R}$ namely correspond to the atoms in the lattice of clopen subsets of $Spec\underline{R}$. We consider the sentences

$\zeta_0 := \forall x\forall y\exists z\ x\neq y \Rightarrow t(x,y,z)=x\ \&\ t(x,z,y)\neq x\ \&\ x\neq z$

$\zeta := \forall x\forall y\exists u\exists v\ (x\neq y \Rightarrow t(x,y,u)=x\ \&\ t(x,u,y)\neq x\ \&\ x\neq u)\ \&$
$\quad\quad \&\ (t(x,v,y)=x\ \&\ t(x,y,v)\neq x)$

5.5 LEMMA : Let $\underline{R}$ be an algebra in the discriminator variety $\mathcal{V}$.

(1) $Spec\underline{R}$ has no isolated points except (possibly) ∇ iff $\underline{R}$ satisfies ζ_0 .

(2) *Spec*$\underline{R}$ has no isolated points at all iff $\underline{R}$ satisfies ζ.

We now turn to the question of determining the special algebras in $\mathcal{W}$, and we want to determine them by the special properties of their spectra and their simple quotients. We first recall some definitions:

Let $\mathcal{K}$ be a class of algebras. An algebra $\underline{I} \in \mathcal{K}$ is called *injective* in $\mathcal{K}$ iff for every embedding $h: \underline{R} \rightarrowtail \underline{S}$ $(R \neq \emptyset)$ in $\mathcal{K}$ and each homomorphism $f: \underline{R} \longrightarrow \underline{I}$ there exists a homomorphism $g: \underline{S} \longrightarrow \underline{I}$ such that $f = g \circ h$.

There are several weaker notions of injectivity which coincide in a variety with CEP (see BANASCHEWSKI [11]), so for $\underline{I} \in \mathcal{W}$ the following are equivalent :

$\underline{I}$ is a *weak injective* , i.e. for every embedding $h: \underline{R} \rightarrowtail \underline{S}$ in and each onto-homomorphism $f: \underline{R} \longrightarrow \underline{I}$ there exists $g: \underline{S} \longrightarrow \underline{I}$ such that $f = g \circ h$

$\underline{I}$ is an *absolute subretract* , i.e. for every embedding $h: \underline{I} \rightarrowtail \underline{S}$ there exists $g: \underline{S} \longrightarrow \underline{I}$ such that $id_I = g \circ h$

$\underline{I}$ has *no proper essential extension* , i.e. every essential embedding $h: \underline{I} \rightarrowtail \underline{S}$ is an isomorphism.

An algebra $\underline{I} \in \mathcal{K}$ is called:
algebraically closed iff every embedding $h: \underline{I} \rightarrowtail \underline{S}$ is pure.
existentially closed iff every embedding $h: \underline{I} \rightarrowtail \underline{S}$ is existential.
If the existentially closed members of $\mathcal{K}$ are first-order definable, then they form the *model-companion* of $\mathcal{K}$, and $\mathcal{K}$ does not have a model companion otherwise. The model-companion is called the *model-completion* of $\mathcal{K}$ iff $\mathcal{K}$ has the amalgamation property.
At the end of this chapter we also deal with *equationally compact* algebras, that are,algebras in which every set of equations with constants from the algebra has a solution provided its finite subsystems have solutions.

In many cases we are able to reduce the problem of determining the special algebras in $\mathcal{W}$ to the much simpler problem of determining the special algebras in the class $\mathcal{A}$ of all simple algebras in $\mathcal{W}$, and in the class $\mathcal{R}$ of relatively complemented distributive lattices with smallest element. For $\mathcal{A}$ the answers are easy to give and for $\mathcal{R}$ they can be derived from the well known results for Boolean algebras (see WĘGLORZ [168]) .

In the class $\mathcal{R}$ it is then clear that the equationally compact algebras are precisely the complete Boolean algebras and that the algebraically closed algebras are atomless and without a greatest element.
For the class $\mathfrak{A}$ the situation is as follows:

<u>5.6 LEMMA :</u> Let $\mathfrak{A}$ be the class of all simple algebras (including the singleton algebra) in the discriminator variety $\mathcal{W}$.

(1) For $\underline{A},\underline{B}\in\mathfrak{A}$, $\underline{A}$ finite, an embedding $f:\underline{A}\rightarrowtail\underline{B}$ is
- (a) essential except in the case $|A| = 1$ & $|B| > 1$.
- (b) pure iff $|A| = 1$ or f is an isomorphism.
- (c) existential iff f is an isomorphism.

(2) T.f.a.e. for $\underline{B}\in\mathfrak{A}$:
- (i) $\underline{B}$ is injective in $\mathfrak{A}$
- (ii) $\underline{B}$ is demi-semi-primal and for every $\underline{A}\in\mathfrak{A}$ either $\underline{A}$ is embeddable into $\underline{B}$ or $\underline{A}$ and $\underline{B}$ have only trivial (= empty or singleton) subalgebras in common
- (iii) $\underline{B}$ is injective in $\mathcal{W}$

(3) T.f.a.e. for $\underline{B}\in\mathfrak{A}$, $|B| \neq 1$:
- (i) $\underline{B}$ is weak injective in $\mathfrak{A}$
- (ii) $\underline{B}$ has no proper extension in $\mathfrak{A}$
- (iii) $\underline{B}$ is finite and algebraically closed in $\mathfrak{A}$
- (iv) $\underline{B}$ is finite and existentially closed in $\mathfrak{A}$
- (v) $\underline{B}$ is weak injective in $\mathcal{W}$

(4) $\underline{B}$ is equationally compact iff $\underline{B}$ is finite.

<u>Proof:</u> (1a) is trivial as all $\underline{A}\in\mathfrak{A}$ are simple. (1,b,c) is trivial for $|\underline{A}| = 1$, so let $A = \{a_1,\dots,a_n\}$ $n\geq 2$ and let $\phi(a_1,\dots,a_n)$ be the positive diagram of $\underline{A}$. As $\underline{A}$ is simple, elements $b_1,\dots,b_n\in\underline{B}$ satisfy $\phi(b_1,\dots,b_n)$ iff they form a singleton algebra or an algebra isomorphic to $\underline{A}$. Let $\bar\phi(a_1,\dots,a_n)$ be the sentence $\exists x\ \phi(a_1,\dots,a_n)\ \&\ \mathop{\&}_{i=1}^{n}(t(a_i,x,a_1) = t(a_i,x,a_2))$.
Then $b_1,\dots,b_n\in\underline{B}$ satisfy $\bar\phi(b_1,\dots,b_n)$ iff they form a singleton algebra or a proper subalgebra of $\underline{B}$ isomorphic to $\underline{A}$. So if f is a proper embedding it can't be pure and neither existential.

(2) (i)<=>(ii): Let $h:\underline{C}\rightarrowtail\underline{D}$ be an embedding in $\mathfrak{A}$ and $f:\underline{C}\longrightarrow\underline{B}$ a homomorphism. (i) says there exists a $g:\underline{D}\longrightarrow\underline{B}$ such

that $f = h\circ g$ and (ii) demands the existence of such a g only in the case $|fC|>1$, but the case $|fC|=1$ is trivial.
(i)=>(iii): Let $h:\underline{R}\rightarrowtail\underline{S}$ be an embedding in $\mathfrak{V}$ and $f:\underline{R}\longrightarrow\underline{B}$ a homomorphism. fR as a subalgebra of $\underline{B}$ is simple, so by CEP there exist maximal congruences Φ on $\underline{R}$ and Ψ on $\underline{S}$ such that $fR\cong\underline{R}/\Phi$ and we have a commuting diagram

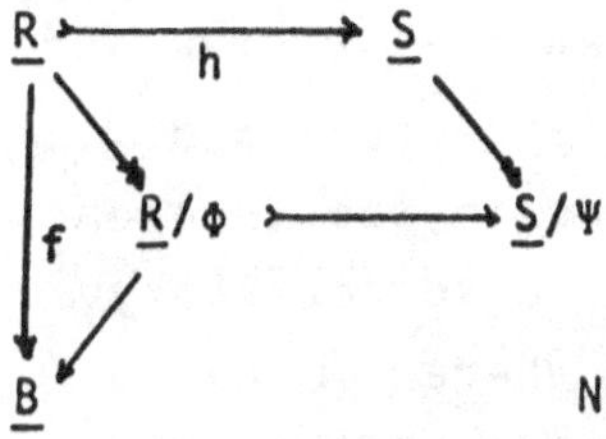

Now the injectivity of $\underline{B}$ within $\mathfrak{A}$ gives the desired diagram completion $g:\underline{S}\longrightarrow\underline{B}$. (iii)=>(i) is trivial.
(3):(v)=>(i)=>(ii)=>(iv)=>(iii) are trivial, and (i)=>(v) is similar to (2,i)=>(2,iii). (ii)=>(i) holds as $\mathfrak{A}$ has the CEP, and (iii)=>(ii) follows from (1,b).
(4): As infinite algebras have proper pure (even elementary) extensions (ultrapowers namely) which don't retract, as they are simple, the finite algebras in $\mathfrak{A}$ are precisely the pure-subretracts (=equationally compact algebras). □

5.7 THEOREM: Let $\mathfrak{V}$ be a discriminator variety, $\mathfrak{A}$ its class of simple algebras, and $\underline{R}\in\mathfrak{V}$ such that $\underline{R}/\Phi$ is algebraically closed in $\mathfrak{A}$ for each $\Phi\in Spec\underline{R}\setminus\{\nabla\}$.

(1) $\underline{R}$ is algebraically closed in $\mathfrak{V}$

(2) $\underline{R}$ is existentially closed in $\mathfrak{V}$ provided one of the following conditions holds

   (a) No non-singleton algebra in $\mathfrak{A}$ has a singleton subalgebra and $Spec\underline{R}$ has no isolated points except ∇.

   (b) Some non-singleton algebra in $\mathfrak{A}$ has a singleton subalgebra and $Spec\underline{R}$ has no isolated points at all

Proof: (1): Let $f:\underline{R}\rightarrowtail\underline{S}$ be an embedding. As all $\underline{R}/\Phi$ are algebraically closed we have for every $\Psi\in Spec\underline{S}$ that $f_\Psi:\underline{R}/f^+\Psi\rightarrowtail\underline{S}/\Psi$ is pure, so by 5.2 f is pure.
(2,a): Let us first remark that under these assumptions on $\mathfrak{V}$ ∇ is always an isolated point in $Spec\underline{R}$, and for any embedding $f:\underline{R}\rightarrowtail\underline{S}$ and $\Psi\in Spec\underline{S}$ we have $f^+\Psi = \nabla \Longleftrightarrow \Psi = \nabla$.
For all $\Psi\in Spec\underline{S}\setminus\{\nabla\}$ $\underline{R}/f^+\Psi$ is algebraically closed and a non-

singleton algebra, hence $\underline{R}/f^+\psi$ is existentially closed and thus $f_\psi:\underline{R}/f^+\psi \rightarrowtail \underline{S}/\psi$ is existential. Now by 5.4 we conclude that f is existential. (2,b): If ∇ is not isolated in $Spec\underline{R}$, then for every embedding $f:\underline{R}\rightarrowtail\underline{S}$ the set $\{\psi\in Spec\underline{S} \mid f^+\psi \neq \nabla\}$ is dense in $Spec\underline{S}\setminus\{\nabla\}$, as $f^+:Spec\underline{S} \twoheadrightarrow Spec\underline{R}$ is a quotient map. So as above we can use 5.4 to conclude that f is existential. □

This theorem only gives a sufficient condition for an algebra $\underline{R}\in\mathcal{V}$ to be algebraically closed or existentially closed, and it is still open, whether the sufficient conditions in the theorem are also necessary. We conjecture that they are also necessary, but we are only able to prove this in the case where $\mathfrak{A}$ has a model-companion $\mathfrak{A}_0\subseteq\mathfrak{A}$. To see this we first note that a model-companion $\mathfrak{A}_0$ of $\mathfrak{A}$ can be $\forall\exists$-axiomatized (see CHANG-KEISLER [25,p111]).

<u>5.8 LEMMA :</u> Let $\mathcal{V}$ be a discriminator variety, $\mathfrak{A}$ its class of simple algebras, and Σ a set of $\forall\exists$-sentences in the language of $\mathfrak{A}$. Then there is a set Σ' of $\forall\exists$-atomic sentences in the language of $\mathcal{V}$ such that for $\underline{R}\in\mathcal{V}$:

$$\underline{R} \models \Sigma' \iff \forall\psi\in Spec\underline{R}\setminus\{\nabla\}\ \underline{R}/\psi \models \Sigma .$$

<u>Proof</u>: For simplicity we can assume that in every sentence $\sigma\in\Sigma$ the only negations that occur are negations of atomic formulas hence inequalities. Now replace in σ each inequality $p(x_1,\ldots,x_n)\neq q(x_1,\ldots,x_n)$ by the equation $t(p(x_1,..),q(x_1,..),x)= =t(p(x_1,..),q(x_1,..),y)$, where x,y are new variables not occuring in σ, and call this new formula σ_1. $\sigma_2:=\forall x\forall y\ \sigma_1$. σ_2 is an $\forall\exists$-sentence , positive, and $\underline{A}\in\mathfrak{A}$ satisfies σ_2 iff $\underline{A}$ satisfies σ or $\underline{A}$ is a singleton. Now use 1.10(4) to replace the $\forall\exists$-positive sentence σ_2 by an $\forall\exists$-atomic sentence σ' . $\Sigma':=\{\sigma' \mid \sigma\in\Sigma\}$. It is now obvious that $\underline{R}\in\mathcal{V}$ satisfies Σ' iff its non-singleton simple quotients satisfy Σ . □

<u>5.9 COROLLARY :</u> Let $\mathcal{V}$ be a discriminator variety and $\mathfrak{A}$ its class of simple algebras. Let $\mathfrak{A}$ have a model-companion $\mathfrak{A}_0\subseteq\mathfrak{A}$ which is axiomatized by the set Σ of $\forall\exists$-sentences. Let Σ' be defined as in 5.8 , and let ζ,ζ_0 be as in 5.5 .

(1) $\underline{R}\in\mathcal{V}$ is algebraically closed iff $\underline{R} \models \Sigma$

(2) a) If no non-singleton algebra in $\mathfrak{A}$ has a singleton subalgebra then $\underline{R}\in\mathcal{V}$ is existentially closed iff $\underline{R} \models \Sigma\cup\{\zeta_0\}$

b) If a non-singleton algebra in $\mathfrak{A}$ has a singleton subalgebra then $\underline{R}\in\mathcal{V}$ is existentially closed iff $\underline{R} \models \Sigma\cup\{\zeta\}$

(3) $\mathcal{V}$ has a model-companion $\mathcal{V}_0\subseteq\mathcal{V}$ and this model-companion is a model-completion iff $\mathfrak{A}_0$ is the model-completion of $\mathfrak{A}$

Proof: Every $\underline{R}\in\mathcal{V}$ can be embedded into an algebra $\underline{S}\in\mathcal{V}$ satisfying Σ' as every $\underline{A}\in\mathfrak{A}$ can be embedded into an algebra $\underline{B}\in\mathfrak{A}_0$. Moreover, depending on whether non-singleton algebras in $\mathfrak{A}$ can have singleton subalgebras or not, $\underline{S}$ can even be chosen to satisfy ζ_0 or ζ. If $\underline{R}$ is algebraically closed this embedding is pure, hence $\underline{R}$ satisfies Σ. If $\underline{R}$ is existentially closed, then the embedding is existential and therefore $\underline{R}$ also satisfies ζ_0 or ζ respectively. Condition (3) follows immediately from 2.5, as $\mathcal{V}$ has the AP iff $\mathfrak{A}$ does. □

In many cases $\mathfrak{A}$ will automatically have a model-companion, if $\mathfrak{A}$ has only finite members e.g. ; the model-companion $\mathfrak{A}_0$ of $\mathfrak{A}$ consists of all maximal members of $\mathfrak{A}$ with respect to the embeddability relation in this case.

Let us now turn to the injectives and weak injectives in a discriminator variety $\mathcal{V}$. We have seen that in $\mathfrak{A}$ the weak injectives are finite and they are precisely those algebras, which are not embeddable into any other (nonisomorphic) member of $\mathfrak{A}$. In order to be injective an algebra $\underline{A}\in\mathfrak{A}$ must be in addition demi-semi-primal and have no non-singleton subalgebra in common with any algebra in $\mathfrak{A}$ which is not embeddable into $\underline{A}$. Using the fact that products and retracts of injective algebras are again injective (BANASCHEWSKI [11]) and that subdirect retracts of weak injectives are again weak injectives (GRÄTZER-LAKSER [61]) we have

5.10 THEOREM : Let $\mathcal{V}$ be a discriminator variety, $\mathfrak{A}$ its class of simple algebras. Let $\underline{A}_i\in\mathfrak{A}$ $(i\in I)$, and let $\underline{R}$ be a retract of $\underline{A}:=\Pi\{\underline{A}_i \mid i\in I\}$ with the inclusion map $f:\underline{R}\rightarrowtail\underline{A}$.

(1) $\underline{R}$ is injective in $\mathcal{V}$ provideo all $\underline{A}_i$ $(i\in I)$ are injective in $\mathfrak{A}$.

(2) $\underline{R}$ is weak injective in $\mathcal{V}$ provided all $\underline{A}_i$ $(i\in I)$ are weak injective in $\mathcal{A}$, and for all projections $p_i:\underline{A}\twoheadrightarrow\underline{A}_i$, the homomorphism $p_i\circ f:\underline{R}\longrightarrow\underline{A}_i$ is onto.

The equationally compact algebras fit nicely into this context because the equationally compact algebras are precisely the pure-injective algebras (see BANASCHEWSKI [11]), so we get similarly

5.11 THEOREM : Let $\mathcal{V}$ be a discriminator variety, and $\mathcal{A}_f$ its class of finite simple algebras. Let $\underline{A}_i\in\mathcal{A}_f$ $(i\in I)$ and $\underline{R}$ a retract of $\Pi\{\underline{A}_i\,|\,i\in I\}$.
Then $\underline{R}$ is equationally compact .

These two theorems give sufficient conditions for algebras $\underline{R}\in\mathcal{V}$ to be injective, weak injective or pure-injective, but it is open whether these conditions are also necessary. In the last part of this section we shall see that these conditions are in fact necessary in residually small discriminator varieties.

For the rest of this section we assume that $\mathcal{V}$ is a residually small discriminator variety and that $\mathcal{A}$ is the set of isomorphism types of simple algebras in $\mathcal{V}$.

Note that $\mathcal{V}$ is definable by a single identity $\varepsilon(x_1,\dots,x_n)$ (see MAKKAI [102] and PADMANABHAN-QUACKENBUSH [111]) and thus $\mathcal{A}$ is defined by the universal positive sentence

$\alpha := \forall x_1\dots\forall x_n\ \varepsilon(x_1,\dots,x_n)\ \&\ (x_1=x_2\ \text{ or }\ t(x_1,x_2,x_3)=x_1\ )$.

For each algebra $\underline{A}\in\mathcal{A}$ we have the $\forall\exists$-sentence

$\delta_A := \forall x\forall y\,\exists x_1\dots\exists x_m\ \phi_A(x_1,\dots,x_m)\&\ \&\{t(x_i,x_j,x)=t(x_i,x_j,y)\,|\,i<j\le m\}$

where ϕ_A is the open positive diagram of $\underline{A}$ and $m = |A|$.
A non-singleton algebra on which t is the discriminator satisfies δ_A iff it contains a subalgebra isomorphic to $\underline{A}$.
Let δ denote the disjunction of all δ_A where $\underline{A}$ is maximal in $\mathcal{A}$ with respect to the embeddability relation, then the $\forall\exists$-sentence $\sigma = \alpha\&\delta$ axiomatizes the set $\bar{\mathcal{A}}$ of all maximal algebras in $\mathcal{A}$ up to isomorphism .

We are now able to characterize the algebraically closed and the existentially closed members of $\mathcal{V}$ and thereby determine the model-companion of $\mathcal{V}$.

5.12 COROLLARY : Let $\mathcal{W}, \mathfrak{A}, \delta$ be as above and ζ, ζ_0 as in 5.5 .

(1) For $\underline{R} \in \mathcal{W}$ the following are equivalent :
  (i) $\underline{R}$ is algebraically closed
  (ii) $\underline{R} \models \delta$
  (iii) Every non-singleton simple quotient of $\underline{R}$ is maximal in $\mathfrak{A}$.

(2) For $\underline{R} \in \mathcal{W}$ the following are equivalent :
  (i) $\underline{R}$ is existentially closed
  (ii) $\underline{R} \models \begin{cases} \delta \& \zeta_0 & \text{if the singleton } 1 \text{ is maximal in } \mathfrak{A} \\ \delta \& \zeta & \text{otherwise} \end{cases}$
  (iii) Every non-singleton simple quotient of $\underline{R}$ is maximal in $\mathfrak{A}$ and *Spec*$\underline{R}$ has no isolated points - except ∇ if 1 is maximal in $\mathfrak{A}$.

(3) $\mathcal{W}$ has a model-companion. The model-companion is a model-completion provided $\mathcal{W}$ is generated by an independent set of demi-semi-primal algebras.

In residually small discriminator varieties we also can fully characterize injectives, weak injectives, equationally compact algebras, and even topologically compact algebras.

5.13 COROLLARY : Let $\mathcal{W}, \mathfrak{A}$, and $\bar{\mathfrak{A}}$ be as above.

(1) $\underline{R} \in \mathcal{W}$ is injective iff $\underline{R} \cong \underline{A}_1[B_1] \times \ldots \times \underline{A}_n[B_n]$, where $B_1, \ldots, B_n$ are complete Boolean algebras, $\underline{A}_1, \ldots, \underline{A}_n \in \bar{\mathfrak{A}}$ all demi-semi-primal and neither of the algebras $\underline{A}_1, \ldots, \underline{A}_n$ has a non-singleton subalgebra in common with any other member of $\bar{\mathfrak{A}}$.

(2) $\mathcal{W}$ has enough injectives iff $\bar{\mathfrak{A}}$ is an independent set of demi-semi-primal algebras

(3) $\underline{R} \in \mathcal{W}$ is weakly injective iff $\underline{R} \cong \underline{A}_1[B_1] \times \ldots \times \underline{A}_n[B_n]$, where $\underline{A}_1, \ldots, \underline{A}_n \in \bar{\mathfrak{A}}$ and $B_1, \ldots, B_n$ are complete Boolean algebras.

(4) $\underline{R} \in \mathcal{W}$ is equationally compact iff $\underline{R} \cong \underline{A}_1[B_1] \times \ldots \times \underline{A}_n[B_n]$, where $\underline{A}_1, \ldots, \underline{A}_n \in \mathfrak{A}$ and $B_1, \ldots, B_n$ are complete Boolean algebras.

(5) $\underline{R} \in \mathcal{W}$ is a compact topological algebra iff $\underline{R} \cong \underline{A}_1^{I_1} \ldots \underline{A}_n^{I_n}$ where $\underline{A}_1, \ldots, \underline{A}_n \in \mathfrak{A}$ and $I_1, \ldots, I_n$ are arbitrary sets.

REMARK: The isomorphism in condition (5) is an algebraic one and it is still open whether a topological algebra in $\mathcal{W}$ can carry a compact Hausdorff topology different from the product topology given by the representation in (5) .

<u>Proof</u>: We start with the weakest assumption that $\underline{R}$ is equationally compact. By Taylor's theorem 5.3 the canonical embedding $\underline{R} \rightarrowtail \Pi\{\underline{R}/\Phi \mid \Phi\in Spec\underline{R}\}$ is pure, so it retracts. Thus $\underline{R}$ is a subdirect retract of the algebras $\underline{R}/\Phi$ $(\Phi\in Spec\underline{R}\setminus\{\nabla\})$ and so by 4.13(1) all the sets $X_{\{A\}}$ $(\underline{A}\in\mathfrak{A})$ are clopen. As these are finitely many the sets $M_A := \{\Phi\in Spec\underline{R} \mid \underline{R}/\Phi \cong \underline{A}\}$ $(\underline{A}\in\mathfrak{A})$ are all clopen and hence $\underline{R} \cong \Pi\{\Gamma\underline{C}(M_A,\underline{A}) \mid \underline{A}\in\mathfrak{A}\} = \Pi\{\underline{A}[M_A^*] \mid \underline{A}\in\mathfrak{A}\}$ (we can omit all factors $\underline{A}\in\mathfrak{A}$ with $M_A=\emptyset$). M_A^* is a complete Boolean algebra by 4.13(2) so (4) is proven.

If $\underline{R}$ is weakly injective all $\underline{A}\in\mathfrak{A}$ with $M_A\neq\emptyset$ have actually to lie in $\overline{\mathfrak{A}}$ because if $\underline{A} \le \underline{B}$ then $\underline{A}[M_A^*] \le \underline{B}[M_A^*]$ and both have the same spectrum, so they can't be retracts of each other unless $\underline{A} \cong \underline{B}$. This proves (3). Similarly one deduces (1) and from (1) immediately (2) follows.

To see (5) we observe that the Boolean algebra $\mathcal{D}(\underline{R})$ is a homomorphic image of $\underline{R}$ by 4.12 , so a compact topology on $\underline{R}$ induces a compact topology on $\mathcal{D}(\underline{R})$. But the only compact Boolean algebras are those of the form $\mathfrak{2}^I$ for some set I (see KAPLANSKY [32]), so $\underline{A}[\mathfrak{2}^I] \cong \underline{A}^I$ proves (5) . □

We conclude this section with some remarks about affine complete algebras in a discriminator variety. We are not able to give a full characterization of the affine complete algebras (not even in the residually finite case) but we are able to give a large class of examples. The striking fact is that all injective, weak injective, and equationally compact algebras are affine complete,and it would be interesting to know, whether this is true also in other varieties.

<u>PROBLEM</u> : In which varieties of algebras are all equationally compact algebras affine complete ?

<u>5.14 AFFINE COMPLETE ALGEBRAS</u>

(1) If $\underline{A}$ is quasi-primal and B is a Boolean algebra, then the Boolean power $\underline{A}[B]$ is affine complete.

(2) If $\underline{R},\underline{S}$ are algebras in the discriminator variety $\mathcal{V}$ and both have compact ∇ , then $\underline{R}\times\underline{S}$ is a <u>*diagonal product*</u> ,

i.e. there is an algebraic function $\perp$ on $\underline{R}\times\underline{S}$ such that $(r,s)\perp(\bar{r},\bar{s}) = (r,\bar{s})$ (see WERNER [171,173]).

(3) If $\underline{A}_1,\dots,\underline{A}_n$ are weakly independent quasi-primal algebras and $B_1,\dots,B_n$ are Boolean algebras, then $\underline{A}_1[B_1]\times\dots\times\underline{A}_n[B_n]$ is affine complete.

(4) In a residually small discriminator variety each equationally compact algebra is affine complete ; in particular each finite algebra is affine complete.

Proof: (1) is a weaker version of theorem7.1 in KEIMEL-WERNER [88]. Let $f:\underline{A}[B]^n \longrightarrow \underline{A}[B]$ be an operation which preserves all congruences on $\underline{A}[B]$. Let X be the Boolean space B^* and $\underline{A}[B]=\Gamma\underset{\sim}{C}(X,\underline{A})$ For each $x\in X$ we have the natural projection $p_x:\underline{A}[B]\longrightarrow\underline{A}_x$, where $\underline{A}_x := \Gamma\underset{\sim}{C}(\{x\},\underline{A}) = \underline{A}$. As f preserves all congruences on $\underline{A}[B]$ it in particular preserves the kernel of p_x . This says there is a unique operation $f_x:\underline{A}^n\longrightarrow\underline{A}$ which makes the diagram

$$\begin{array}{ccc} \underline{A}[B]^n & \xrightarrow{\quad f \quad} & \underline{A}[B] \\ \downarrow & & \downarrow p_x \\ \underline{A}^n & \overset{f_x}{\dashrightarrow} & \underline{A} \end{array}$$

commutative. As $\underline{A}$ is finite, for each mapping $g:\underline{A}^n \longrightarrow \underline{A}$ the set $X_g:=\{x\in X\,|\,f_x=g\}$ is clopen in X . $\underline{A}$ has only finitely many n-ary operations, and they are all algebraic functions as $\underline{A}$ is functionally complete. Let $\{g_1,\dots,g_m\}$ be the set of all n-ary operations on $\underline{A}$. On $\Gamma\underset{\sim}{C}(X_g,\underline{A}) = \underline{A}[X_g^*]$ the algebraic function g agrees with f thus the operation $(\dots(g_1\perp g_2)\perp\dots)\perp g_m$ agrees with f on the algebra

$(\dots(\underline{A}[X^*_{g_1}]\times\underline{A}[X^*_{g_2}])\times\dots)\times\underline{A}[X^*_{g_n}]\cong\ \underline{A}[B]$, hence f is an algebraic function on $\underline{A}[B]$.

(2): As in $\underline{R}$ and $\underline{S}$ ∇ is principal, we have elements $a,\bar{a}\in R$ and $b,\bar{b}\in S$ such that for all $x\in R$ $t(a,\bar{a},x)=a$ and for all $y\in S$ $t(b,\bar{b},y)=b$. Define $(r,s)\perp(\bar{r},\bar{s}) :=$ $t(t((a,b),(a,\bar{b}),(r,s)),(a,b),t((a,b),(\bar{a},b),(\bar{r},\bar{s})))$ this equals $(t(t(a,a,r),a,t(a,\bar{a},\bar{r})),t(t(b,\bar{b},s),b,t(b,b,\bar{s}))) = (r,\bar{s})$.

(3) By WERNER [171] the diagonal product of affine complete algebras is affine complete, so the result follows from (1),(2) .

(4) follows from (3) by 5.13 . □

We should remark at this point that not all affine complete algebras necessarily have the form as in (3), as ISKANDER [68] has examples of affine complete algebras in which the universal congruence ∇ is not compact. The above theorem however covers the theorem of HU [65], that in a variety generated by independent primal algebras all algebras are affine complete, this is true because in these varieties all algebras are of the form as in (3) . It also covers the theorem 7.1 in [88] and PIXLEY's result [119], that in a discriminator variety each finite algebra is affine complete. There is, however, a much better version of this result due to BAKER-PIXLEY [9] stating the same for each arithmetical variety (see 1.17) and this is not covered by the above theorem.

§ 6 DECIDABILITY

Interesting first order theories which are decidable are few in number (see ERSHOV et al. [41,p 94]), and if we consider the first-order theories of varieties, then the list is essentially

(1) Boolean algebras (TARSKI [161], 1949)
(2) Abelian groups (SZMIELEW [153], 1954)
(3) mono-unary algebras (EHRENFEUCHT [38] , 1959)
(4) relatively complemented distributive lattices (ERSHOV [40], 1964)
(5) n-valued Post-algebras (ERSHOV [39], 1967)
(6) $(x^n = x)$-rings (COMER [31],1973)
(7) residually finite varieties of monadic algebras (COMER [32], 1974).

As is well known (1),(2), and (4) were proved by quantifier elimination (introduced by SKOLEM in 1919). Ershov proved (5) by combining (1) with FOSTER's [45] Boolean power representation of Post-algebras. Ehrenfeucht used semi-models for the proof of (3). (6) and (7) were proved by Comer using a representation of the countable models by subsheaves of constant sheaves (for (6) this representation is due to AHRENS-KAPLANSKY [3]) adapting the fundamental lemma of FEFERMAN-VAUGHT [43] to such sheaves and then using the following theorem of RABIN [129] ,(1969) :

6.1 THEOREM (RABIN)

The first-order theory, with quantification over ideals, of countable Boolean algebras is decidable.

A common feature of all the examples above, except (2) and (3), is that these varieties are generated by a weakly independent set of quasi-primal algebras. For these varieties we can use the same technique as Comer did to get a decidability result, and we can extend our list of decidable theories considerably.
We are going to prove the following

6.2 THEOREM : Let $\mathcal{V}$ be a residually small discriminator variety.
The first-order theory, with quantification over congruences, of the countable members of $\mathcal{V}$ is decidable.

6.3 COROLLARY :

Every residually small discriminator variety has a decidable first-order theory

This corollary follows immediately from the theorem 6.2 by the LÖWENHEIM-SKOLEM-theorem.
As new varieties with decidable theory we have e.g.

(8) residually small varieties of dually-Stone Heyting-algebras (2.6)

(9) Łukasiewicz-algebras of order n (2.8)

(10) residually small varieties of n-dimensional cylindric algebras (2.9)

(11) residually small varieties of relation algebras (2.10)

(12) residually small varieties of biregular rings (2.12)

(13) residually small varieties of Baer*-rings (2.12)

(14) varieties of arithmetical rings (2.11), they are the same as residually small varieties of strongly regular rings (2.12) and they are the subvarieties of (6)

(15) Commutative complementary semigroups of rank n (2.13)

The proof of theorem 6.2 proceeds in three major steps (6.4 - 6.6). For the finite set $\mathfrak{A}$ of simple $\mathcal{V}$-algebras we construct in 6.4 a finite partial algebra $\underline{P}$ which contains isomorphic copies of the members of $\mathfrak{A}$ as subalgebras, and we show in 6.5 that each countable member of $\mathfrak{V}$ has a representation by a subsheaf of $\underset{\sim}{C}(Spec\underline{R},\underline{P})$, and so we assign to each $\underline{R}\in\mathfrak{V}$ the Boolean algebra $(Spec\underline{R})^*$ together with the filters $\hat{X}_F$ of all clopen sets which contain the closed subset X_F of $Spec\underline{R}$ which is the set of all $\Phi\in Spec\underline{R}$ such that the stalk over Φ in the subsheaf of $\underset{\sim}{C}(Spec\underline{R},\underline{P})$ is contained in the subalgebra F of $\underline{P}$. In 6.6 we give a translation of each sentence in the language of $\underline{R}$ into a sentence in the language of $(Spec\underline{R})^*$ with filter predicates $\hat{X}_F$ for certain subalgebras F of $\underline{P}$ and thereby reduce the decidability of $Th(\mathfrak{V})$ to Rabin's theorem.

6.4 THE CONSTRUCTION OF P

Let $\mathfrak{V}$ be a residually small discriminator variety and $\mathfrak{A}$ the set of isomorphism types of simple $\mathfrak{V}$-algebras, including the singleton algebra. $\mathfrak{A}$ is a finite set of finite algebras and partially ordered by the embeddability relation $\leq$. Now by order-induction we construct for every $\underline{A}\in\mathfrak{A}$ a partial algebra $\underline{P}_A$, all of whose complete subalgebras are isomorphic to members of $\mathfrak{A}$.
If $\underline{A}$ is $\leq$-maximal we define $\underline{P}_A := \underline{A}$.
If $\underline{A}$ has the upper neighbors $\underline{A}_1,\dots,\underline{A}_n$ we let $\underline{Q}_A$ be the disjoint union of $\underline{P}_{A_1},\dots,\underline{P}_{A_n}$. Let E_A denote the set of all embeddings $f:\underline{A}\rightarrowtail\underline{Q}_A$ and $\underline{Q}_A\times E_A$ the disjoint union of $|E_A|$ many copies of $\underline{Q}_A$. On this partial algebra the relation $(x,f)\sim(y,g)$ iff $x=y$ & $f=g$ or $\forall a\in A\ x=fa$ & $y=ga$ is a congruence relation on $\underline{Q}_A\times E_A$ and we let $\underline{P}_A$ be the quotient algebra $\underline{Q}_A\times E_A/\sim$.
The important property of the algebra $\underline{P}_A$ is the following

(1) $\underline{P}_A$ is an amalgamation of $\underline{Q}_A$ over $\underline{A}$.

Obviously we get the same result if we replace E_A by a system of representatives of E_A modulo the equivalence induced by the automorphisms of $\underline{Q}_A$.
If $\underline{A}_1,\dots,\underline{A}_n$ are the minimal elements of $\mathfrak{A}$, we let $\underline{P}$ be the disjoint union of the partial algebras $\underline{P}_{A_1},\dots,\underline{P}_{A_n}$.
We let F be the system of complete subalgebras of $\underline{P}$; note that each $F\in F$ is isomorphic to some member of $\mathfrak{A}$.
Observe that in the case where $\mathfrak{A}$ is the set of subalgebras of

a single demi-semi-primal algebra $\underline{A}$, $\underline{A}$ itself can serve as the algebra $\underline{P}$, provided the automorphisms of $\underline{A}$ are transitive on the one-element subalgebras ; in the other case $\underline{P}$ is the disjoint union of the isomorphism types of $\underline{A}$ with a distinguished singleton-subalgebra , and glued together over the distinguished one-element subalgebras.

6.5 THE COUNTABLE $\mathfrak{W}$-ALGEBRAS

It is quite obvious that a countable algebra $\underline{R}\in\mathfrak{W}$ has a spectrum $Spec\underline{R}$ with a countable basis, and we will see that $Spec\underline{R}$ then has the disjoint refinement property. From this we are going to show that the canonical sheaf of $\underline{R}$ is embeddable as a subsheaf into the constant sheaf $\underset{\sim}{C}(Spec\underline{R},\underline{P})$.

(1) A Boolean space with a countable basis has the DRP .

Proof: Let $\{N_i | i<\omega\}$ be a basis for the topology on the Boolean space X , let $U\subseteq X$ be open and $\{U_t | t\in T\}$ an open cover of U . W.l.o.g. we can assume all N_i $(i<\omega)$ to be clopen. We now define a disjoint refinement $\{V_j | j<\omega\}$ of $\{U_t | t\in T\}$ recursively:

$V_j := N_{i_j} \setminus \bigcup\{V_k | k<j\}$ where $i_j := \min\{i | i>i_{j-1}$ & $\exists t\ N_i \subseteq U_t\}$. □

(2) If $\underline{R}\in\mathfrak{W}$ is countable, the standard sheaf $\underset{\sim}{S}(\underline{R})$ is embeddable as a subsheaf into $\underset{\sim}{C}(Spec\underline{R},\underline{P})$.

Proof: By 3.9 we have to show that $\underset{\sim}{S}(\underline{R})$ is locally embeddable into $\underset{\sim}{C}(Spec\underline{R},\underline{P})$. Let $\Phi\in Spec\underline{R}$ and $\underline{R}/\Phi \cong \underline{A}\in\mathfrak{A}$, then we find sections σ_a $(a\in A)$ such that tha map $a\longmapsto\sigma_a(\Phi)$ is an isomorphism $\underline{A}\longrightarrow\underline{R}/\Phi$, and thus by 3.8(1) there is a clopen neighborhood N of Φ such that for each $\Psi\in N$ the map $a\longmapsto\sigma_a(\Psi)$ is an embedding $\underline{A}\rightarrowtail\underline{R}/\Psi$.
We show by order-induction on $\underline{A}\in\mathfrak{A}$ that $\underset{\sim}{S}(\underline{R})|N$ is embeddable into $\underset{\sim}{C}(N,\underline{P}_A)$. If $\underline{A}$ is maximal in $\mathfrak{A}$ then all mappings $a\longmapsto\sigma_a(\Psi)$ $(\Psi\in N)$ are isomorphisms and hence $\underset{\sim}{S}(\underline{R})|N \cong \underset{\sim}{C}(N,\underline{A}) = \underset{\sim}{C}(N,\underline{P}_A)$.
Let $\underline{A}$ have the upper neighbors $\underline{A}_1,\dots,\underline{A}_n$ in $\mathfrak{A}$ and let $U = \{\Phi\in N |$ some $\underline{A}_i$ is embeddable into $\underline{R}/\Phi\}$. By 3.8(2) U is open and by induction hypothesis $\underset{\sim}{S}(\underline{R})|U$ is locally embeddable into $\underset{\sim}{C}(U,\underline{Q}_A)$ as $\underline{Q}_A$ contains all algebras $\underline{P}_{A_1},\dots,\underline{P}_{A_n}$. As $Spec\underline{R}$ has the DRP by (1), we can apply 3.9(5) to get a global embedding $\underset{\sim}{S}(\underline{R})|U \hookrightarrow \underset{\sim}{C}(U,\underline{Q}_A)$. $\underline{P}_A$ is an amalgamation of $\underline{Q}_A$

over $\underline{A}$, so we can apply 3.10 to obtain an embedding $\underset{\sim}{S}(\underline{R})|N \hookrightarrow \underset{\sim}{C}(N,\underline{P}_A)$. As $\underline{P}$ contains all algebras $\underline{P}_A$ ($\underline{A}\in \mathfrak{A}$) the claim (2) is proved. □

Now we have for each countable $\underline{R}\in \mathfrak{M}$ the Boolean space $Spec\underline{R}$ with the countable basis $(Spec\underline{R})^*$ and an embedding γ: $\underset{\sim}{S}(\underline{R}) \longrightarrow \underset{\sim}{C}(Spec\underline{R},\underline{P})$ which defines for each subalgebra $F\in F$ of $\underline{P}$ a closed subset X_F of $Spec\underline{R}$
$X_F := \{\Phi\in Spec\underline{R} \mid$ the stalk over Φ of $\gamma(\underset{\sim}{S}(\underline{R}))$ is contained in $F\}$
and by 3.12 we have that $\underline{R}$ is isomorphic to the algebra of continuous mappings $f:Spec\underline{R} \longrightarrow \underline{P}$ such that for each $F\in F$ $f(X_F) \subseteq F$. Moreover we have $Spec\underline{R} = \bigcup\{X_F \mid F\in F\}$. The closed sets X_F corresponds to the filter $\mathfrak{X}_F$ of all clopen sets of $Spec\underline{R}$ which contain X_F.

(3) The mapping $r \mapsto (r^{-1}(p) \mid p\in P)$ is a bijection between the continuous mappings $r:Spec\underline{R} \longrightarrow \underline{P}$ and the P-tuples $(x_p \mid p\in P)$ in the Boolean algebra $B = (Spec\underline{R})^*$ satisfying $\bigvee\{x_p \mid p\in P\} = 1$ and $\&\{x_p \wedge x_q = 0 \mid p,q\in P, p\neq q\}$.
Moreover $r:Spec\underline{R} \longrightarrow \underline{P}$ satisfies $r(X_F)\subseteq F$ iff $\bigvee\{x_p \mid p\in F\} \in \mathfrak{X}_F$.

Using the galois correspondence between the closed sets of $Spec\underline{R}$ and the congruences of $\underline{R}$ we define for each $\Theta\in C(\underline{R})$ $\hat{\Theta}$ to be the filter of all clopen sets of $Spec\underline{R}$ which contain the closed set $V\Theta$.

(4) Let $r,s\in R$ and let $(x_p \mid p\in P),(y_p \mid p\in P)$ be the P-tuples corresponding to r,s according to (3).
For $\Theta\in C(\underline{R})$ we have $(r,s)\in\Theta$ iff $\bigvee\{x_p \wedge y_p \mid p\in P\}\in\hat{\Theta}$

Proof: $\bigvee\{x_p \wedge y_p \mid p\in P\}$ is just the set $E(r,s)$ under the interpretation (3), so $(r,s)\in\Theta$ iff $E(r,s) \supseteq V\Theta$ proves (4). □

6.6 THE TRANSLATION

Let $L(\mathfrak{M})$ be the language of $\mathfrak{M}$ with quantification over congruences and let $L(\mathfrak{B})$ be the language of Boolean algebras with quantification over filters. Small letters are used for first-order variables, $X,\ldots$ for a congruence variable and $\mathfrak{X},\ldots$ for a filter variable.

We define a translation $T: L(\mathcal{W}) \longrightarrow L(\mathcal{B})$ recursively as follows

(i) (atomic formulas)

$$x = r \longmapsto \&\{x_p = r^{-1}(p) \mid p \in P\}$$

$$f(x_1,\dots,x_n) = y \longmapsto \underset{p\in P}{\&} \quad y_p = \bigvee\{\bigwedge_{i\le n} x_{ip_i} \mid p_i \in P, f(p_1,\dots,p_n)=p\}$$

$$(x,y)\in\Theta \longmapsto \bigvee_{p\in P} (x_p \wedge y_p) \in \hat{\Theta}$$

$$(x,y)\in X \longmapsto \bigvee_{p\in P} (x_p \wedge y_p) \in \hat{X}$$

(ii) (Sheffer stroke)

$$\alpha \mid \beta \longmapsto T\alpha \mid T\beta$$

(iii) ($\exists$-quantifier)

define: $\mathrm{Adm}(x_p \mid p\in P) := \underset{p\neq q}{\&}\, x_p \wedge x_q = 0 \;\&\; \bigvee_{p\in P} x_p = 1 \;\&\; \underset{F\in F}{\&}\; \bigvee_{p\in F} x_p \in \hat{X}_F$

$$\exists x\ \alpha \longmapsto (\exists x_p)_{p\in P}\ T\alpha \ \&\ \mathrm{Adm}(x_p \mid p\in P)$$

$$\exists X\ \alpha \longmapsto \exists X\quad T\alpha$$

(1) Let $\underline{R}\in\mathcal{W}$ be a countable algebra, $B=(Spec\underline{R})^*$, and X_F $F\in F$ as above. For each sentence $\alpha\in L(\mathcal{W})$ we have:

$$\underline{R} \models \alpha \quad \text{iff} \quad (B,(X_F)_{F\in F}) \models T\alpha$$

This statement is easily proved by induction on the length of α and an immediate corollary from that is the following:

(2) Let $\mathcal{W}_0$ be the class of countable $\mathcal{W}$-algebras and $\mathcal{B}_0$ the class of countable Boolean algebras and let $\alpha\in L(\mathcal{W})$.

$$\mathcal{W}_0 \models \alpha \text{ iff } \mathcal{B}_0 \models (\exists X_F)_{F\in F}((\forall x(\underset{F\in F}{\&} x\in X_F) \Rightarrow x=1) \ \&\ T\alpha\ )$$

To see this observe that $\forall x(\underset{F\in F}{\&} x\in X_F) \Rightarrow x=1$ says that $Spec\underline{R}$ is the only clopen set containing all X_F $(F\in F)$ which, because of the finiteness of F, is equivalent to $Spec\underline{R} = \bigcup\{X_F \mid F\in F\}$. Claim (2) now is the reduction of the decision problem of $\mathcal{W}_0$ to the countable Boolean algebras (with quantification over filters) which finishes the proof of theorem 6.2 . □

A striking fact about our list of varieties with decidable theory is that all of these varieties are residually small, and it would be interesting to know if there is a general principle behind it.

PROBLEM Is every variety with a decidable first-order theory residually small ?

CONJECTURE For discriminator varieties the answer is yes .

We want to give some evidence for this conjecture.
In a recent abstract in the *Notices* of the AMS 22(1975),75T-E63, RUBIN claims that the theory of monadic algebras has an undecidable theory. We haven't seen the proof yet, but if he is correct then each variety of cylindric algebras of finite dimension which is not residually small has an undecidable theory, because it has the variety of monadic algebras as a reduct. That the variety of cylindric algebras of dimension $n \geq 2$ has an undecidable theory was shown in COMER [29] , in fact he shows that $\mathcal{C}_n$ has even a recursively inseparable theory.

For the class $\mathcal{R}e$ of relation algebras TARSKI [160] has shown that even the equational theory of $\mathcal{R}e$ is undecidable and it is not difficult to see that each subvariety of $\mathcal{R}e$ which contains all finite simple representable relation algebras (= the algebras of all relations on a finite set) has an undecidable theory too.

The variety $\mathcal{H}_+$ of all dually-Stone Heyting-algebras has an undecidable theory too, because the theory of finite distributive lattices, which by ERSHOV-TAITSLIN [42] is undecidable, can be interpreted in the theory of $\mathcal{H}_+$. The finite simple members of $\mathcal{H}_+$ are just all finite distributive lattices with a new greatest element adjoined. For the same reason each subvariety of $\mathcal{H}_+$ which contains all finite subdirectly irreducible members of $\mathcal{H}_+$ also has an undecidable theory.

For all varieties which do have to do with rings, we can combine the results of TARSKI et al. [159] and ROBINSON [130] to see that each subvariety of $\mathcal{RB}^*$ or $\mathcal{RS}$, which contains rings of characteristic 0 (and therefore the rationals) , has an undecidable theory. Using TAITSLIN [158] one also gets that the variety of all Baer*-rings of characteristic p also has an undecidable theory.

R e f e r e n c e s

[1] Adams, D.H. : Equational classes of Foulis-semigroups and orthomodular lattices (Preprint 1973)

[2] Adams, D.H. : Semigroups with no non-zero nilpotent elements
Math. Z. 123(1971), 168 - 176

[3] Ahrens, R.F.
Kaplansky, I. : Topological representation of algebras
Trans. AMS 63 (1948), 457 - 481

[4] Ash, C.J. : Reduced powers and Boolean extensions
J.London Math.Soc. 9 (1975),429 -432

[5] Bacsich, P.D. : Primality and model completions
Alg. Univ. 3 (1973), 265 - 270

[6] Bairamov, R.A. : On the question of functional completeness in many-valued logic
Disc. Analiz 11 (1967), 3 - 20

[7] Bairamov, R.A. : Stabilizators of predicates and Sheffer-functions in finite-valued logics
Doklady Akad.Nauk 24,2 (1968),3 - 6

[8] Bairamov, R.A. : Sheffer functions in many-valued logics
Problems in the theory of electronic digital computers 3(1967), 62 - 71

[9] Baker, K.A.
Pixley, A.F. : Polynomial interpolation and the chinese remainder theorem for algebraic systems
Math. Z. 143 (1975), 165 - 174

[10] Balbes, R.
Dwinger, Ph. : Distributive lattices
Univ. of Missouri Press, Columbia(1974)

[11] Banaschewski,B. : Injectivity and essential extensions in equational classes of algebras

Proc. Conf. Universal Alg. Kingston Ontario (1969), 131 - 147

[12] Bell, J.L.
Slomson, A.B. : Models and Ultraproducts
North Holland, Amsterdam (1969)

[13] Berberian, S.K. : Baer*-rings
Springer, Berlin,Heidelberg, New York
Grundlehren 195(1972)

[14] Birkhoff, G. : Lattice Theory
AMS Coll.Publ. 25(1967), 3rd edition

[15] Bodnarcuk, V.G.
Kaluznin, L.A.
Kotov, V.N.
Romov, B.A. : Galois theory for Post algebras
Cybernetics 5 (1969), 243-252, 531-539

[16] Bosbach, B. : Komplementäre Halbgruppen, ein Beitrag zur Idealtheorie kommutativer Halbgruppen
Math. Ann. 161 (1965), 279 - 295

[17] Bosbach, B. : Komplementäre Halbgruppen, eine Darstellungstheorie
Math. Ann. 179 (1968), 1 - 14

[18] Bosbach, B. : Komplementäre Halbgruppen, Axiomatik und Arithmetik
Fund. Math. 64 (1969), 257 - 287

[19] Bosbach, B. : Komplementäre Halbgruppen, Kongruenzen und Quotienten
Fund. Math. 69 (1970), 1 - 14

[20] Bosbach, B. : Rechtskomplementäre Halbgruppen, Axiome, Polynome, Kongruenzen
Math Z. 124 (1972), 273 - 288

[21] Bulman-Fleming, S.
Werner, H. : Equational compactness in quasi-primal varieties (Preprint 1975)

[22] Burris, S. : Boolean Powers (Preprint 1974)

[23] Burris, S. : An example concerning definable principal congruences (preprint 1974)

[24] Burris, S.
Werner, H. : Sheaf constructions and their elementary properties, I, II (preprint 1975)

[25] Chang, C.C.
Keisler, H.J. : Model Theory
North Holland, Amsterdam, (1973)

[26] Chin, L.H.
Tarski, A. : Distributive and modular laws in relation algebras
U. of Cal. Publ.Math. 1(1951), 341 - 384

[27] Clark, D.M.
Krauss, P.H. : Para-primal algebras, (Preprint 1975)

[28] Comer, S.D. : Complete and model complete theories of monadic algebras (Preprint 1974)

[29] Comer, S.D. : Finite inseparability of some theories of cylindrification algebras
J. Symb. Logic 34 (1969), 171 - 176

[30] Comer, S.D. : Representations by algebras of sections over Boolean spaces
Pacific J. Math. 38 (1971), 29 - 38

[31] Comer, S.D. : Elementary properties of structures of sections
Bol. Soc. Math. Mexicana (to appear)

[32] Comer, S. D. : Monadic algebras with finite degree (Preprint 1974)

[33] Dauns, J.
Hofmann, K.H. : The representation of biregular rings by sheaves
Math. Z. 91 (1966), 103 - 123

[34] Davey, B.A. : m-Stone lattices
Can. J. Math. 24 (1972), 1027 - 1032

[35] Davey, B.A. : Sheaf spaces and sheaves of universal algebras
Math. Z. 134 (1973), 275 - 290

[36] Davis, G. : Rings with orthogonality relations
Bull. Austral. Math.Soc. 4(1971),163-178

[37] Day, A. : A note on the congruence extension property
Alg. Univ. 1 (1971), 234 - 235

[38] Ehrenfeucht,A. : Decidability of the theory of one function
Notices AMS 6 (1959), 268

[39] Ershov, Yu.L. : On the elementary theory of Post-varieties
Algebra i Logika 6 (1967), 7 - 15

[40] Ershov, Yu.L. : Decidability of the elementary theory relatively complemented distributive

lattices and the theory of filters
Algebra i Logika 3 (1964), 17 -38

[41] Ershov, Yu.L.
Lavrov, I.A.
Taimanov, A.D.
Taitslin, M.A. : Elementary theories
Russ. Math. Surveys 20 (1965),35 - 105

[42] Ershov, Yu.L.
Taitslin, M.A. : The undecidability of certain theories
Algebra i Logika 2 (1963), 37 -42

[43] Feferman, S.
Vaught, R.L. : The first-order properties of products of algebraic systems
Fund. Math. 47 (1959),57 - 103

[44] Fleischer, I : A note on subdirect products
Acta Math.Sci.Hungar. 6 (1955),463-465

[45] Foster, A.L. : Generalized Boolean theory of universal algebras I, II
I Math.Z. 58 (1953), 306 - 336
II Math.Z. 59 (1953), 191 - 199

[46] Foster, A.L. : The identities of - and unique subdirect factorization whithin - classes of universal algebras
Proc. AMS 7 (1956), 1011 - 1013

[47] Foster, A.L. : An existence theorem for functionally complete universal algebras
Math. Z. 71 (1959), 69 -82

[48] Foster, A.L. : Generalized equational maximality and primal-in-the-small algebras

[49] Foster, A.L. : Functional completeness in the small I,II
I Math. Ann. 143 (1961), 29 - 53
II Math. Ann. 149 (1962), 173 - 191

[50] Foster, A.L. : Families of algebras with unique (sub-) direct factorization
Math. Ann. 166 (1966), 302 - 326

[51] Foster, A.L. : Semi-primal algebras, characterization and normal decomposition
Math. Z. 99 (1967), 105 - 116

[52] Foster, A.L. : Automorphisms and functional completeness in universal algebra
Math. Ann. 180 (1969), 138 - 169

[53] Foster, A.L. : Congruence relations and functional completeness in universal algebra I, II
I Math. Z. 113 (1970), 293 - 308
II Math. Z. 115 (1970), 23 - 32

[54] Foster, A.L. : Functional completeness and Automorphisms I, II
I Monatsh. f. Math. 75 (1971), 303 - 315
II Monatsh. f. Math. 76 (1972), 226 - 238

[55] Foster, A.L., Pixley, A.F. : Semi-categorical algebras I, II
I Math. Z. 83 (1964), 147 -169
II Math. Z. 85 (1964), 169 -184

[56] Foster, A.L., Pixley, A.F. : Algebraic and equational semi-maximality, Equational spectra I, II
I Math. Z. 92 (1966), 30 - 50
II Math. Z. 94 (1966), 122 - 133

[57] Foster, A.L., Pixley, A.F. : Total algebras and weak independence I,II
I Math. Z. 123 (1971), 93 - 104
II Math. Z. 125 (1972), 271 - 284

[58] Gould, M.I., Grätzer, G. : Boolean extensions and normal subdirect powers of finite universal algebras
Math. Z. 99 (1967), 16 - 25

[59] Grätzer, G. : Universal algebra
Van Nostrand, Princeton (1968)

[60] Grätzer, G. : On the class of subdirect powers of a finite algebra
Acta Sci. Math. (Szeged) 25 (1964),160-168

[61] Grätzer, G., Lakser, H. : The structure of pseudocomplemented distributive lattices I, II, III
I Trans. AMS 156 (1971), 335 - 342
II Trans. AMS 156 (1971), 343 - 358
III Trans. AMS 169 (1972), 475 - 487

[62] Henkin, L., Monk, J.D., Tarski, A. : Cylindric algebras I
North Holland, Amsterdam (1971)

[63] Hofmann, K.H. : Representations of algebras by continuous sections
Bull. AMS 78 (1972), 291 - 373

[64] Hu, T.K. : On the fundamental subdirect factorization theorems of primal algebra theory
Math. Z. 112 (1969), 154 - 162

[65] Hu, T.K. : Characterization of algebraic functions in equational classes generated by independent primal algebras
Alg. Univ. 1 (1971), 187 - 191

[66] Hu, T.K. : On the topological duality for primal algebra theory
Alg. Univ. 1 (1971), 152 - 154

[67] Hu, T.K. : Locally equational classes of universal algebras (preprint)

[68] Iskander, A.A. : Algebraic functions on p-rings
Coll. Math. 25 (1972), 37 - 41

[69] Iskander, A.A. : On Boolean extensions of primal algebras
Math. Z. 124 (1972), 203 - 207

[70] Jablonskiĭ, S.V. : On functional completeness in the three-valued calculus
Dokl. Akad. Nauk 95 (1954), 1153 -1155

[71] Jablonskiĭ, S.V. : Functional constructions in the k-valued logic
Trudy Math.Inst. Steklova 51(1958),5-142

[72] Jablonskiĭ, S.V.
Gavrilov, G.P.
Kudrjavcev, V.B. : Functions of the algebra of logic and the post-classes
(Russian) Moskow (1966)

[73] Jõnsson, B. : Algebras whose congruence lattices are distributive
Math. Scand. 21 (1967), 110 - 121

[74] Jõnsson, B. : Algebraic structures with prescribed automorphism groups
Coll. Math. 19 (1968), 625

[75] Jõnsson, B.
Tarski, A. : Representation problems for relation algebras
Bull. AMS 54 (1948), 80, 1192

[76] Jõnsson, B.
Tarski, A. : Boolean algebras with operators I, II
I Amer. J. Math. 73 (1951), 891 - 939
II Amer. J. Math. 74 (1952), 123 - 162

[77] Kaiser, H.K. : On a problem in the theory of primal algebras (Prdprint)

[78] Kaiser, H.K. : A class of locally complete universal algebras (Preprint)

[79] Kaiser, H.K. : Über lokal polynomvollständige universelle Algebren (Preprint)

[80] Kaiser, H.K. : Über das Interpolationsproblem in nichtkommutativen Ringen (Preprint)

[81] Kaiser, H.K : Der Defekt abelscher Gruppen
Monatsh. f. Math. 78 (1974), 109 - 116

[82] Kaplansky, I. : Topological rings
Amer. J. Math. 69 (1947), 153 - 183

[83] Keimel, K. : Darstellungen von Halbgruppen und universellen Algebren durch Schnitte in Garben, bireguläre Halbgruppen
Math. Nachr. 45 (1970), 81 - 96

[84] Keimel, K. : The representation of lattice ordered groups and rings by sections in sheaves
Springer Lecture Notes Math. 248 (1972)

[85] Keimel, K. : Baer extensions of rings and Stone extensions of semigroups
Semigroup Forum 2 (1971), 55 - 63

[86] Keimel, K. : Anneaux rêticulês quasi-rêguliers et hyperarchimêdiens
C.R.Acad. Sci. Paris 266(1968),A524-A525

[87] Keimel,K. : Reprêsentation d'anneaux rêticulês dans des faisceaux
C.R.Acad.Sci.Paris 266 (1968),A124-A127

[88] Keimel, K.
Werner, H. : Stone duality for varieties generated by quasi-primal algebras
Memoirs AMS 148 (1975), 59 - 85

[89] Kist, J. : Compact spaces of minimal prime ideals
Math. Z. 111, (1969), 151 - 158

[90] Knoebel, R.A. : Simplicity versus functional completeness
Math. Ann. 189 (1970), 299 - 307

[91] Knoebel, R.A. : Primal extensions of universal algebras
Math. Z. 115 (1970), 53 - 57

[92] Krauss, P. : On primal algebras
Alg. Univ. 2 (1972), 62 - 67

[93] Krauss, P. : On quasi-primal algebras
Math. Z. 134 (1973), 85 - 89

[94] Lakser, H. : The structure of pseudocomplemented distributive lattices
Trans. AMS 156 (1971), 335 - 342

[95] Lausch, H.
Nöbauer, W. : Algebra of polynomials
North Holland, Amsterdam (1973)

[96] Lavrov, I.A. : Effective inseparability of the sets of identically true formulae and finitely refutable formulae for certain elementary theories
Algebra i Logika 2(1963), 5 - 18

[97] Lindsey, D.
Madison, B. : Bounded Baer-Levi Semigroups(Preprint)

[98] Lipshitz, L.
Saracino, D. : The model companion of the theory of commutative rings without nilpotent elements
Proc. AMS 38(1973), 381 - 387

[99] Łukasiewicz, J. : O logike trojwartosciowiej
Ruch Filozoficzny 5 (1920)

[100] Lyndon, R.C. : The representation of relation algebras
Amer. J. Math. 51 (1950), 707 - 729

[101] Macintyre, A. : Model completeness for sheaves of structures
Fund. Math. 81 (1973), 73 - 89

[102] Makkai, M. : A proof of Baker's finite-base theorem on equational classes generated by finite elements of congruence distributive varieties
Alg. Univ. 3 (1973), 174 - 181

[103] Mal'cev, A.I. : On the general theory of algebraic systems
Math. Sbor. (77) 35 (1954), 3 - 20

[104] Maurer, W.D.
Rhodes, G.L. : A property of finite simple non-abelian groups
Proc. AMS 16 (1965), 552 - 554

[105] Mewborn, A.C. : Some conditions on semiprime rings
J. Alg. 13 (1969), 422 - 431

[106] Michler, G.
Wille, R. : Die primitiven Klassen arithmetischer Ringe
Math. Z. 113 (1970), 369 - 372

[107] Monk, D. : On equational classes of algebraic versions of logic
Math. Scand. 27 (1970), 53 - 71

[108] Mulvey, C.J. : Représentation des produits sous-direct d'anneaux per espaces annelés
C.R.Acad.Sci.Paris 270(1970),A564-A567

[109] Neumann, W.D. : On Mal'cev conditions
Austr. J. Math. 17 (1974), 376 - 388

[110] Padmanabhan, R.
Quackenbush, R.W. : Equational theories of algebras with distributive congruences
Proc. AMS (to appear)

[111] Peercy, D.E. : Complexes and the complete Baer extension of a commutative ring
Math. Z. (to appear)

[112] Pierce, R.S. : Modules over commutative regular rings
Memoirs AMS 70 (1967)

[113] Pixley, A.F. : Distributivity and permutability of congruence relations in equational classes of algebras
Proc. AMS 14 (1963), 105 - 109

[114] Pixley, A.F. : Functionally complete algebras generating distributive and permutable classes
Math. Z. 114 (1970), 361 - 372

[115] Pixley, A.F. : The ternary discriminator function in universal algebra
Math. Ann. 191 (1971), 167 - 180

[116] Pixley, A.F. : Local Mal'cev conditions
Can. Math. Bull. 15 (1972), 559 - 568

[117] Pixley, A.F. : A note on hemi-primal algebras
Math. Z. 124(1972), 213 - 214

[118] Pixley, A.F. : Local weak independence and primal algebra theory
Boll. U. Math. Ital. (4) 5 (1972), 381 - 399

[119] Pixley, A.F. : Completeness in arithmetical algebras
Alg. Univ. 2 (1972), 17z - 196

[120] Pixley, A.F. : Equationally semi-complete varieties
Alg. Univ. 4 (1974), 323 - 327

[121] Pöschel, R. : Postsche Algebren von Funktionen über einer Familie endlicher Mengen
Z.math.Logik Grundl.Math. 19(1973),37-74

[122] Pöschel, R. : Die funktionale Vollständigkeit von Funktionenklassen über einer Familie endlicher Mengen
Z.math.Logik Grundl.Math. 20(1974),537-550

[123] Post, E.L. : Introduction to a general theory of elementary propositions
Amer. J. Math. 43 (1921), 163 - 185

[124] Quackenbush, R.W. : Equational classes generated by finite algebras
Alg. Univ. 1 (1971), 265 - 266

[125] Quackenbush, R.W. : Demi-semi-primal algebras and Mal'cev-type conditions
Math. Z. 122(1971), 166 - 176

[126] Quackenbush, R.W. : The triangle is functionally complete
Alg. Univ. 2 (1972), 128

[127] Quackenbush, R.W. : Structure theory for equational classes generated by quasi-primal algebras
Trans. AMS 187 (1974), 127 - 144

[128] Quackenbush, R.W. : Semi-simple equational classes with distributive congruence lattices
(Preprint 1974) .

[129] Rabin, M.O. : Decidability of second order theories and automata on infinite trees
Trans. AMS 141 (1969), 1 - 35

[130] Robinson, G. : The decision problem for fields
Model Theory, North Holland, Amsterdam

[131] Rosenberg, I. : La structure des fonctions de plusieurs variables sur un ensemble fini
C.R.Acad.Sci.Paris 260(1965),3817-3819

[132] Rosenberg, I. : Zu einigen Fragen der Superposition von Funktionen mehrerer Veränderlicher
Bull. Inst. Polyt. 12 (1966), 1-2

[133] Rosenberg, I. : Über die Verschiedenheit maximaler Klassen in P_k
Rev. Roum. Math. 3(XIV)(1969),431 - 438

[134] Rosenberg, I. : Complete sets for finite algebras
Math. Nachr. 44(1970), 253 - 258

[135] Rosenberg, I. : Über die funktionale Vollständigkeit in den mehrwertigen Logiken
Rozpravy Cesk.Akad. 80 (1970)

[136] Rosenberg, I. : Algebren und Relationen
EIK 6 (1970) 2, 115 - 124

[137] Rosenberg, I. : A classification of universal algebras by infinitary relations
Alg. Univ. 1 (1971), 350 - 354

[138] Rosenberg, I. : Strongly rigid relations
Rocky Mountaan J.Math 3(1973),631 - 639

[139] Rosenberg, I. : Universal algebras with all operations of bounded range
Coll. Math. 30 (1974), 177 - 185

[140] Rosenberg, I. : Some maximal closed classes of operations on infinite sets
Math. Ann. 212 (1974), 157 - 164

[141] Rosenberg, I. : A semilattice on the set of permutations on an infinite set
Math. Nachr. 60 (1974), 191 - 199

[142] Rosenberg, I. : A Galois-connection between universal algebras and relations (Preprint 1975)

[143] Rosenberg, I. : On generating large classes of Sheffer functions (Preprint 1975)

[144] Rosenberg, I. : Functional completeness in heterogeneous multiple-valued logics (Preprint 1975)

[145] Rosenberg, I. : Special types of universal algebras preserving a relation (preprint 1975)

[146] Rousseau, G. : Completeness in finite algebras with a single operation
Proc. AMS 18 (1967), 1009 - 1013

[147] Saracino, D. : Model companions for $\aleph_0$-categorical theories
Proc. AMS 39 (1973), 591 - 598

[148] Shelah, S. : Every two elementarily equivalent models have isomorphic ultrapowers
Israel J. Math. 10 (1971), 224 - 233

[149] Sheffer, H.M. : A set of five independent postulates for Boolean algebras with applications to

logical constants
Trans. AMS 14 (1913), 481 - 488

[150] Śierpinski, W. : Sur les fonctions des plusieurs variables
Fund. Math. 33 (1945), 169 - 173

[151] Sioson, F.M. : Some primal clusters
Math. Z. 75 (1961), 201 - 210

[152] Sioson, F.M. : Free algebra characterization of primal and independent algebras
Proc. AMS 12 (1961), 435 - 439

[153] Szmielew, W. : Elementary properties of abelian groups
Fund. Math. 41 (1954), 203 - 271

[154] Schweigert, D. : Über endliche ordnungspolynomvollständige Verbände
Monatsh. f. Math. 78 (1974), 68 - 76

[155] Stone, M.G. : Subalgebra and automorphism structure in universal algebra (Preprint 1973)

[156] Stone, M.H. : Application of the theory of Boolean rings to general topology
Trans. AMS 41 (1937), 375 -481

[157] Stone, M.H. : The theory of representations of Boolean algebras
Trans. AMS 40 (1936), 37 - 111

[158] Taitslin, M.A. : Undecidability of the elementary theories of certain classes of finite commutative associative rings
Algebra i Logika 2 (1963), 29 - 51

[159] Tarski, A.
Mostowski, A.
Robinson, R.M. : Undecidable theories
North Holland, Amsterdam, (1968)

[160] Tarski, A. : An undecidable system of sentential calculus
J. symb. logic 18 (1953), 189

[161] Tarski, A. : Arithmetical classes and types of Boolean algebras
Bull. AMS 55 (1949), 63 - 64

[162] Tarski, A. : On the calculus of relations
J. symb. logic 6 (1941), 73 - 89

[163] Tarski, A. : Relation algebras (mimeographed notes)

[164] Taylor, W. : Characterizing Mal'cev conditions
(Preprint 1973)

[165] Taylor, W. : Residually small varieties
Alg. Univ. 2 (1972), 33 - 53

[166] Webb, D.L. : Generation of any n-valued logic by one binary operation
Proc. Nat. Acad. Sc. 21(1935), 252-259

[167] Webb, D.L. : Definition of Post's generalized negative and maximum in terms of one binary operation
Amer. J. Math. 51 (1936), 193 - 194

[168] Węglorz, B. : Equationally compact algebras I
Fund. Math. 59 (1966), 289 - 298

[169] Werner, H.
Wille, R. : Charakterisierungen der primitiven Klassen arithmetischer Ringe
Math. Z. 115 (1970), 197 - 200

[170] Werner, H. : Eine Charakterisierung funktional vollständiger Algebren
Archiv d. Math. 21 (1970), 383 - 385

[171] Werner, H. : Produkte von Kongruenzklassengeometrien universeller Algebren
Math. Z. 121 (1971), 111 - 140

[172] Werner, H. : A Mal'cev condition for admissible relations
Alg. Univ. 3 (1973), 263

[173] Werner, H. : Diagonal products and affine completeness
Alg. Univ. 4 (1974), 269 - 270

[174] Werner, H. : Finite simple non-abelian groups are functionally complete
Bull.Soc.Roy.Sci.Liège 43(1974), 400

[175] Werner, H. : Congruences on products of algebras and functionally complete algebras
Alg. Univ. 4 (1974), 99 - 105

[176] Werner, H. : Which partition lattices are congruence lattices?
Coll.Math.Soc.János Bolyai (to appear)

[177] Werner, H. : Varieties generated by quasi-primal algebras have decidable theories
Coll.Math.Soc.János Bolyai (to appear)

[178] Wille, R. : Kongruenzklassengeometrien
Springer Lecture Notes Math. 113 (1973)

[179] Wille, R. : Allgemeine Algebra-zwischen Grundlagenforschung und Anwendbarkeit. (Preprint)

[180] Wolf, A. : Sheaf representation of arithmetical algebras
Memoirs AMS 148 (1974), 87 - 93

Universale Algebren und Theorie der Radikale

Herausgegeben von Hans-J. Hoehnke

(Studien zur Algebra und ihre Anwendungen)

1976. VI, 85 Seiten — 2 Abb. — gr. 8° — M 18,—

Bestell-Nr. 762 223 7 (2174/1)

Dieser Sammelband ist der Theorie der Lokalisierungen, der Allgemeinen Radikaltheorie und der Allgemeinen Dualitätstheorie gewidmet. Er enthält Original- und Übersichtsarbeiten zu diesen Themen und erfaßt damit die Hauptvorträge der „Internationalen Tagung über Universale Algebren und Theorie der Radikale" in Reinhardsbrunn. Verfasser der Beiträge sind zum größten Teil internationale Spitzenwissenschaftler. Der Inhalt betrifft fundamentale klassische algebraische Theorien, die sowohl in der DDR wie im Ausland weiterentwickelt wurden und von aktueller Bedeutung sind.

Bestellungen durch eine Buchhandlung erbeten

AKADEMIE-VERLAG

DDR-108 Berlin, Leipziger Str. 3—4

Anhand einer Fülle von Beispielen belegt dieses Buch zunächst die Bedeutung des Studiums der quasiprimalen bzw. funktional vollständigen Algebren. Die Arbeit gibt einen Überblick aller bis zur Drucklegung bekannten Ergebnisse in diesem Gebiet. Zentral ist eine Garben-Darstellung der hier betrachteten Algebren, mit deren Hilfe eine Fülle modelltheoretischer Ergebnisse gewonnen werden können, z. B. Charakterisierungen der algebraisch oder existentiell abgeschlossenen Algebren, der injektiven und der gleichungskompakten Algebren, sowie die Entscheidbarkeit der Theorie 1. Stufe einer Varietät, wie sie hier betrachtet wird.
Giving numerous examples this book proves the importance of a study of quasi primal and functionally complete algebras. It presents a review of all results in this field known by its time of print. The central part is a sheaf-representation of the algebras in question. This has a lot of model theoretic consequences like characterizations of the algebraically closed or existentially closed algebras, of the injective and the equationally compact algebras and finally the decidability of the first-order theory of the varieties under consideration.

FSC
www.fsc.org
MIX
Papier | Fördert gute Waldnutzung
FSC® C083411

Kolibri 360 GmbH
Ferdinand-Jühlke-Straße 7
99095 Erfurt, Deutschland
produktsicherheit@kolibri360.de